More praise for *Bl*

A *Science News* Favorite Bo

"[*Blight*] is a short, crisp introduction to the possibility of being devoured by fungi." —Heather Smith, *Sierra*

"A book this endlessly fascinating, by an author whose astonishing zeal for detail makes her knowledge feel bottomless, is the crystal clear kind of science writing we need to face the changes we've wrought on this planet." —Alan Weisman, author of *The World Without Us* and *Countdown*

"Emily Monosson brings eloquence and clarity to her enigmatic subject: the often invisible existence of fungi and the incredible ways it influences and impacts our lives. In turn fascinating and frightening, *Blight* will alarm readers in the best possible way—by changing how they look at the world around them."

—M. R. O'Connor, author of *Wayfinding: The Science and Mystery of How Humans Navigate the World*

"The spores are coming for us. But is blight our plight? Not if Emily Monosson has something to say about it. She shows us that solutions to this vexing problem lie with us. Our iconic human endeavors, from agriculture and gardening to globe-trotting, must accommodate nature's needs. Monosson's ideas can bear the needed fruit—a way of life that is better and safer for people and the many plants and animals on which we depend." —Anne Biklé, coauthor of *What Your Food Ate* and *The Hidden Half of Nature*

"Time and again, our habits of global travel and commerce have let loose fungal pathogens with devastating effect, as Emily Monosson shows in gripping stories of harmful fungi decimating frogs, bats, bananas, American chestnuts, white pines, and other species. These past fights add up to an indelible cautionary tale, making *Blight* required reading for the post-COVID age."

—Susan Freinkel, author of *Plastic: A Toxic Love Story*

"Monosson is a skilled writer, capable of translating complicated scientific topics into compelling layperson's terms . . . she crafts a thrilling narrative." —*Kirkus Reivew*, starred review

"Monosson commendably serves as a medical Paul Revere by persuasively warning us that dangerous fungi are already causing havoc. . . . Neglecting these emerging organisms is truly hazardous to health."

—*Booklist*, starred review

"Although this is an expert's account, grounded in exhaustive interviews with researchers at the forefront of mycology, the language is clear and plain. It's an urgent message meant for a wide audience, and the tone is by turns alarming and reassuring."

—Jennifer Latson, *Boston Globe*

"Emily Monosson presents an eye-opening, and at times grisly, account of fungal diseases that threaten pine trees, bananas, frogs, bats and, increasingly, people." —Erin Garcia de Jesús, *Science News*

"Scientists and others are working . . . to put into place systems and policies to head off a new pandemic that could make Covid look like child's play. In this terrifying book, Monosson gives us the opportunity to wise up and join their efforts."

—Priscilla Long, *American Scholar*

"*Blight* is about fungi, yes, but it's also about facets of modern life that researchers have been worried about for quite some time— globalization, biosecurity, climate change." —Zoya Teirstein, *Grist*

ALSO BY EMILY MONOSSON

Natural Defense:
Enlisting Bugs and Germs to
Protect Our Food and Health

Unnatural Selection:
How We Are Changing
Life, Gene by Gene

Evolution in a Toxic World:
How Life Responds to Chemical Threats

Motherhood, the Elephant in the Laboratory:
Women Scientists Speak Out

BLIGHT

FUNGI AND THE
COMING PANDEMIC

EMILY MONOSSON

W. W. NORTON & COMPANY
Independent Publishers Since 1923

For information about permission to reproduce selections from this book, write to Permissions, W. W. Norton & Company, Inc., 500 Fifth Avenue, New York, NY 10110

For information about special discounts for bulk purchases, please contact W. W. Norton Special Sales at specialsales@wwnorton.com or 800-233-4830

Manufacturing by Lakeside Book Company
Book design by Daniel Lagin
Production manager: Lauren Abbate

Library of Congress Cataloging-in-Publication Data

Names: Monosson, Emily, author.
Title: Blight : fungi and the coming pandemic / Emily Monosson.
Description: First edition. | New York, NY : W. W. Norton & Company, [2023]|
Includes bibliographical references and index.
Identifiers: LCCN 2023009948 | ISBN 9781324007012 (cloth) |
ISBN 9781324007029 (epub)
Subjects: LCSH: Pathogenic fungi—Popular works. | Medical mycology—
Popular works. | Mycoses—Popular works.
Classification: LCC QR245 .M658 2023 | DDC 616.9/6901—dc23/eng/20230504
LC record available at https://lccn.loc.gov/2023009948

ISBN 978-1-324-10518-3 pbk.

W. W. Norton & Company, Inc., 500 Fifth Avenue, New York, N.Y. 10110
www.wwnorton.com

W. W. Norton & Company Ltd., 15 Carlisle Street, London W1D 3BS

10 9 8 7 6 5 4 3 2 1

CONTENTS

Part I
CONSEQUENCES

Part II
RESOLUTION

AUTHOR'S NOTE

For most species discussed in this book, whether fungi, plant, or animal, I use a mixture of Latin names and common names depending on the organism. Most often I will be using a species' common name, but many microbes have only Latin names. Genes for specific proteins, when mentioned, are italicized.

I struggled with whether measurements such as weight or temperature ought to be reported as metric units (since the metric system is used by much of the world and scientists) or imperial units, which we use here in the United States. In the end, since I think most readers are more familiar with miles rather than kilometers, the imperial system won out. The smallest unit of weight in the imperial system is an ounce. Rather than estimating fractions of an ounce, I chose to stick with grams as scientists report in their studies. There are twenty-eight grams in an ounce. When I write about temperature, I will use Celsius and include the conversion to Fahrenheit.

Finally, there are many different government agencies and some nonprofit organizations in these pages, and their names are long. I will either use their whole name or refer to them by their initials when those initials are familiar to readers (such as CDC, NASA, and USDA).

INTRODUCTION

In a large cave on the side of a mountain in Vermont, a fungal spore rests on the mud floor. It has been there since early spring when the bats, mostly little brown bats, began moving out to their summer roosts. The cave had been their winter hibernaculum, a place to hang and conserve energy through the cold when food is scarce. The spore, a microscopic capsule gently curved like a caraway seed or banana, can survive for months resting on the cave floor. In the fall the bats will return. As they settle in for the winter, one, a young female born earlier in the spring, sips some water from a puddle on the floor. She has flown dozens of miles from the roost where she was born to this cave. Her wing brushes the floor where the spore has been resting, and that is all it needs to begin its own journey. Before long the spore, now resting on the bat's membranous wing rather than the muddy floor, germinates and begins to grow. As it does, it sends out threadlike fungal hyphae in search of food.

Bat skin has plenty of what this fungus needs; it thrives on keratin. The cave is cold, a few degrees above freezing, but this particular fungus prefers the cold. It will infiltrate the little brown bat's wings and feed on her until it is time for it to reproduce, and then it will drop hundreds of spores and pieces of hyphae, which will land on other bats, the cave wall, and the mud floor. The young female will not survive the infection. The bits of fungi in time will lead to the death of other bats with whom she shared the cave. Hundreds, then millions of bats will become infected, and most—but not all—will die.

The spore belongs to a fungus called *Pseudogymnoascus destructans*, which causes a disease known as white nose syndrome. The disease has killed millions of bats since its emergence sometime in the 2000s. The epidemic may have begun as I described, with a spore landing on a cave floor.

Meanwhile, in the mountainous western region of the country, spores from a very different fungus called a "rust" fill the air on a cool, moist spring day. Some of these spores will land on the evergreen needles of a whitebark pine. The tree is old and grows twisted from decades of being blasted by wind and harsh weather. It is a survivor. The spore germinates in the damp, misty air. It begins to grow into the needle, the twig, the branch, the tree. The tree has never had to contend with this type of fungus, and whatever natural defenses it has are insufficient. As the fungus grows, needles will brown and die. A few years later the entire tree will die. The scenario has happened millions of times to millions of trees. This fungus, which causes an infection called white pine blister rust, arrived on the continent over a century ago and remains a problem today. The fungus that is killing bats emerged barely two decades ago and is also spreading across the continent. Most scientists say that once a fungus arrives and takes hold, it won't "go away." The rust fungus is here to stay, and so too is the bat fungus. Trees, bats, and other species impacted by fungi are in peril of dying off, lost forever.

Collectively, infectious fungi and fungus-like pathogens are the most devastating disease agents known on the planet. The incidence of novel fungal diseases across species, including humans, has risen over the past century. Where do these fungi come from? And can we do anything to prevent their emergence?

• ■ •

RUSTS, MOLDS, MILDEWS, AND MUSHROOMS—WE LIVE IN A CLOUD of fungal spores, microscopic spherules of nascent fungi. Fungi are everywhere. In the back of the refrigerator a thick mat of green greets us when we open a tub of forgotten yogurt. Pink and yellow fuzz blooms across the wall of a flood-damaged home. The air we breathe is teeming with fungal spores, while belowground tangles of hyphae burrow through soil; water, nutrients, and other chemicals flow through these cellular tubules. There are fungi in the deep ocean, in the radioactive ruins of Chernobyl, and behind damp towels hung to dry on the International Space Station orbiting the planet. If you make sourdough bread, you know that yeasts—which are also fungi—are floating around the kitchen mostly innocuously. Within each of the tiny black dots of a *Rhizopus* fungus we see blooming on a piece of old bread, there are tens of thousands of spores. Their number makes them visible. A fungus called corn smut makes a rounded, container-like gall that can send twenty-five billion spores across a field. Chemists estimate that some fifty million tons of spores populate the atmosphere on an annual basis, each spore carried aloft on the wind from earth-dwelling fungi. Though most fungi rely on spores, not all do. Yeasts and their relatives bud, pinching off daughter cells, although under duress even yeasts will make spores.

Most fungi travel via spores. Some move only fractions of an inch, dropping from one leaf to another. Others cross oceans, surviving low atmospheric temperatures, desiccating air, and the harsh sunlight that comes with high-altitude travel. Humans carry spores too—on our clothes, in our hair—and we breathe them in and out by the thousands. Spores that have landed in the mud might be carried in the recesses of our work boots and sneakers and the treads of our car and truck tires. We drop them along a hiking trail as we move, or we board an airplane, providing the fungus with a lift from one state to another or across an ocean. Fungal spores are aboard the International Space Station; some may travel with us to the Moon or even Mars. Some are hardier than others. There are spores that must germinate quickly or die and so may not survive long and arduous travel, while others survive but die when they land on an unsuitable host. Some lie dormant until conditions are right: perhaps a host sets down roots in a speck of soil where a spore has rested for decades. This is, in part, what makes some fungi so problematic. Some disease-causing fungi can survive in the environment for days or months or years even without a host. Unlike many other pathogenic microbes, fungal spores can be remarkably persistent, and each single spore carries the instructions for a next generation of mold, mildew, smut, or myriad other fungi. By some estimates there are at least six million different species of fungi, most as yet unknown (for perspective, there are roughly two million known animal species, with the majority of those insects, and nearly four hundred thousand land plant species not including mosses, liverworts, and the like). Fungi are some of the most prolific and diverse life-forms on the planet.

Most fungal species are essential for the survival of plants, animals, and humans. Our gut and skin microbiomes are populated not only with bacteria but viruses, protozoans, and fungi too, all of which are members of the dynamic and diverse community of microbes that we carry within and upon us. Belowground, mycorrhizal fungi—fungi that colonize the roots of plants—connect and nourish trees and other plants and help control other microbes that may cause disease. The life-saving penicillin discovered through a combination of serendipity and astute observation by Alexander Fleming is made by the same kind of fungus we might see on a moldering bit of bread or infiltrating the rind of an old cantaloupe.

Many fungi reveal themselves to us as mushrooms, fleshy reproductive bodies with fanciful names like Pink Disco, Destroying Angel, and Dead-Man's Fingers. Morels, chanterelles, hens-of-the-woods, and others are prized in the kitchen. A few make chemicals that can send us on a psychedelic trip or kill us. Mushrooms take shape from the probing, ever-expanding threadlike hyphae that bind together and rise up—pushing through bark, soil, sometimes even blacktop. These and other fungal fruiting structures (because not all fungi form mushrooms) release spores. When they grow, their hyphae lace through the ground and grow in and around the rhizomes of trees and other plants, or they may laze across rotting logs. Fungi are some of the most important decomposers of the world, turning what once lived back into nutrients and soil. Though fungi may remind us of plants, they are in fact more closely related to animals, and like both plants and animals, they belong to their own taxonomic kingdom, the kingdom Fungi.

Animals, even microscopic animals, consume and digest food. We digest our food in our stomach. Fungi digest food first, sending their digestive juices into the environment. Enzymes decompose plants and animals and other microbes. A birch felled by high winds, bathroom wallboard, a block of cheese, a human body are all converted back into nutrients in this way. Fungi that feed on the dead, the so-called saprobes, reduce skin and feather, bark and leaf into their molecular building blocks, amino acids, fatty acids, and simple sugars, which are in turn nutrients for fungi, plants, and other living things. The fungus then absorbs what it needs from this broken-down matter. If it were not for fungi, the world would be piled high with the deceased and would be virtually uninhabitable. Most fungi live if not in collaboration, then in peace with other living things. But some do not. Some feed on the living rather than the dead and dying. Most fungi give life; fungal pathogens take it.

You may have read about the great frog die-off. Or if you live in the northeastern United States, you noticed the disappearance of bats or are familiar with the deadly yeast circulating in hospitals and nursing homes. If you love coffee, cocoa, or bananas, you may have read an article or two with worrisome headlines about their potential demise. Yet fungal threats are not one-off, single-species problem oddities, as news headlines might suggest. They are big, continuing, diverse, and potentially catastrophic. Each species or population we lose is a loss with consequences. Frogs and bats eat moths and insects, and so their loss is an opportunity for crop-eating caterpillars and disease-carrying insects to flourish. Nut and pine trees support whole ecosystems—bears, birds, fish, other plants, and communities of microbes including mushrooms that we might like to eat. What happens when a bear or a bird suddenly loses its food?

I tell these stories of loss through the perspective of the ecologists, foresters, physicians, biologists, policy makers, and citizens who are working against time to save plant, animal, and human life as we know it. Many agree that once a disease-causing fungus settles in, it will not "go away": in other words, a fungus that can persist in the environment without its host is here for the long haul. All agree that a species' best chance for surviving a fungal pandemic is its genetic diversity. Some are trying to prevent future fungal pandemics by tightening trade and travel policy or applying new technologies. Others have spent decades breeding disease-resistant trees, even though they will never live to see the outcome of their work. Crop scientists, geneticists, and seed savers seek resistance in crop diversity to ensure there will be bananas or wheat for the next generation. Their collective efforts offer some glimmer of hope if we as a society can pay attention.

Prevention is difficult but not impossible. It means being more careful about the plants and animals we move around the world; performing rapid disease diagnostics and being willing to respond when a test turns up positive; and, even more, caring about prevention when *we* move around the globe. It also means protecting biodiversity and reducing habitat loss. Where prevention has failed, protecting genetic diversity and the potential resistance genes in a population of trees or salamanders or a food crop is the next best way to survive a fungal pandemic. But there are no guarantees for a genetic rescue. When we remove natural barriers between species, one outcome is the emergence of novel disease. Most interactions will be benign, some will be mildly problematic, others will be catastrophic.

The pandemics and epidemics I write about in these pages all began with a fungus that was moved from its home environment to a completely new setting where it happened upon a suitable host. There are a lot of fungi in the world. Although most are harmless, a few can cause utter devastation when provided with a novel and susceptible host. Our job going forward will be to prevent the potentially harmful fungi from meeting the susceptible host—including us.

PART I

CONSEQUENCES

Chapter 1

EMERGENCE

On November 4, 2016, the US Centers for Disease Control and Prevention (CDC) *Mortality and Morbidity Weekly Report* headlined an unlikely kind of infection. Though previously unrecognized, the infection was turning up in various locations around the world almost simultaneously. It was new, was difficult to diagnose, and had a high mortality rate, killing between 30 to 60 percent of infected patients. Even when diagnosed properly, it resisted medication. The disease could also be transmitted from one patient to another, and it was stubborn, quickly contaminating hospital equipment and rooms. This bizarre infection was caused by *Candida auris*, a fungus that is categorized as a yeast. The emergence of a new fungal pathogen is unusual but not unheard of, particularly among the ever-growing population of immunocompromised patients, including those on powerful steroid drugs, cancer survivors, and transplant recipients. But *C. auris* was also unusual because it emerged with a resistance to some or all antifungal drugs, depending on the strain. It seemed to spread between patients and between hospitals, a characteristic more reminiscent of bacteria and viruses than fungi. And it was unusual because it seemed to come from nowhere and everywhere at once.

The CDC is responsible for, among other things, disease surveillance and identification. Its first mission, dating back to the 1940s, was malaria control, but it has since morphed into the premier agency for tracking and alerting the medical world to baffling new infections, from Ebola to meningitis, influenza, HIV, and COVID-19. The CDC now attracts some of the best physicians, veterinarians, microbiologists, and epidemiologists, among other scientists. In 2015, before *C. auris* became a known problem, a medical mycology laboratory in Pakistan sent some samples to the agency's Fungal Reference Laboratory. There had been an outbreak, and the Pakistani scientists sought confirmation that the fungus was what they had thought it was—a yeast called *Saccharomyces cerevisiae*. The yeast, also known as brewer's yeast, is used in making beer, wine, and breads, but under some circumstances it can infect humans and make us sick. The CDC's laboratory found that the yeast had been misidentified. Rather than *S. cerevisiae* it was instead *C. auris,* which had been identified about a decade earlier, swabbed from the ear of a patient in Japan and named accordingly, *Candida auris*, a fungus of the ear.

A year later *C. auris* began popping up in other locales around the globe, including a handful of cases, diagnosed retrospectively, in the United States, which is when the CDC issued its first warning about the emergent disease. Once the warning went out, other cases were identified. Hundreds of cases in the United States and thousands of cases globally have since been reported. In April 2019 the *New York Times* published a story about a patient who had been hospitalized at Mount Sinai in Brooklyn, New York, with *C. auris*. He died three months later. By then the yeast had colonized the entire room. Dr. Scott Lorin, the hospital's president told the *Times* that "everything was positive—the walls, the bed, the doors, the curtains, the phones, the sink, the whiteboard, the poles, the pump. The mattress, the bed rails, the canister holes, the window shades, the ceiling, everything in the room was positive." Because the yeast resisted disinfectants used at the time, the hospital ripped out tiles from both the floor and ceiling. Tom Chiller is the chief of the CDC's Mycotic Diseases Branch and has decades of experience working with challenging diseases. *C. auris*, he says, is a big concern because no one knows how it emerged or where it came from and because it is so resistant to treatment. He has referred to the fungus as "a creature from the black lagoon."

Even more troubling, scientists worry that the yeast is a harbinger for other human fungal pathogens. Over the past century fungal infections have caused catastrophic losses in other species, but so far we have been lucky. Our luck may be running out.

· ● ·

THE WORLD IS LITTERED WITH YEASTS THAT LIVE ON OR IN FIRE-bugs, apples, rubber trees, flowers, soil, mangroves, blue-striped grunts, and myriad other life-forms. Of the thousands of species categorized as yeasts, including *Candida*, most won't harm us. Some yeasts are even normal members of our gut microbiome—the community of bacteria, viruses, yeasts, and other microbes living in our gut. As with most communities, members have different roles, and microbiomes are no exception: some microbes are beneficial, sharing resources or space with their neighbors; others are competitive, lobbing harmful chemicals to prevent neighboring microbes from consuming their food or moving into their patch of our gut or skin. Some do their own thing and are neither helpful nor harmful. Scientists are only beginning to understand these complex relationships and how these microbes influence us. Most disease-causing microbes are not normal members of the human microbiome but outsiders: an influenza virus; *Borrelia burgdorferi*, the bacterium that causes Lyme disease; or *Salmonella*, which causes food poisoning. But sometimes a typically innocuous microbe like *Candida albicans,* a member of our gut flora that tends to live peacefully with us for much of our lives—when given the right, or wrong, conditions—can make us sick or even kill us.

Both *C. albicans* and *C. auris* are members of the taxonomic category Ascomycota, a sundry group of fungi that includes the eponymous dead-man's fingers, brain-like morels, and delicate scarlet elfin cups. Within the Ascomycota, *Candida* fall into a smaller grouping called *Saccharomycotina*, or yeasts. Some are closely related and others distant. *Candida* yeasts are an odd collection of life that as a group has been described as a "wastebasket genus that includes hundreds of very unrelated species." To emphasize the distance between these microbial relatives, one scientist told me that *C. auris* and *C. albicans* are as different from one another as humans are from fish. What yeasts do have in common is their tendency to exist as single cells like bacteria, though with little or no character, which makes it difficult to discern one species from another unless coaxed into colonies on a petri dish.

When yeasts reproduce by the millions, colonies eventually appear. Most often, yeasts reproduce asexually by budding off clones, doubling their population every one hundred minutes or so, a life cycle more like bacteria than fungi. Rarely, under certain conditions, yeasts behave like more typical fungi and grow hyphae or reproduce sexually. This ability to spawn offspring with or without sex is common to many if not all fungi. Some, like yeast, favor asexual reproduction, while others cycle between sexual and asexual. Fungi are fantastically diverse when it comes to sex and reproduction; at least one fungus has thousands of different "mating types." When yeasts do engage in sex, they stretch their typically round bodies into ghost-like "shmoos," the name borrowed from a mid-twentieth-century *Li'l Abner* cartoon creature. Sexually receptive shmoos are drawn to one another by chemical scents called pheromones. When the yeasts are cultured and reproduce on a petri dish, the invisible cells over time pile up into pinpricks, and pinpricks eventually expand into rounded mounds or flat disks looking much like spattered paint. Colonies of *C. auris* typically appear cream or white in color, but on some media they take on shades of pink or purple.

A fungal infection of the skin, genitals, or throat might cause us to hide our toes or send us to the doctor's office, but it is seldom fatal. This isn't true once a fungus invades our blood. Each year nearly a billion patients struggle with fungal infections; for 150 million it will be life threatening, and of those, more than 1.6 million people around the world will die. This statistic is similar to the number of deaths from tuberculosis and nearly three times higher than deaths from malaria. Some invasive

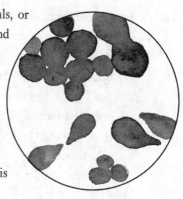

YEAST (AND SHMOOS)

fungi, depending on species, strain, and other conditions, have fatality rates of 30 to 100 percent. In the hospital, *Candida* fungi are a common cause of blood-borne infections, with most caused by *C. albicans*. Even before the emergence of *C. auris*, systemic yeast infections were notorious for their high mortality rate, which hovers around 40 percent, despite antifungal drug treatment.

The first reports of a rise in *Candida* infections from thrush (caused by yeast growing in the mouth and throat) to vaginal infections or, worse, rare fungal infections of major internal organs like the heart began in the 1950s. The cause was antibiotics. Some have even referred to yeast infections as a "disease of antibiotics." For all the benefits of antibiotics (and there are many), the powerful drugs can kill off beneficial bacteria in addition to the targeted bacterial infection. Broad-spectrum antibiotics encourage fungal overgrowth by killing off gut bacteria like *Lactobacillis* and *Bifidobacterium,* which can help keep *Candida* species in check. When the numbers of these bacteria decrease, a fungus like *C. albicans* has an opportunity to explode. Antibiotics can make us more vulnerable to once harmless fungi.

Fortunately, physicians can proactively treat patients who are on long-term antibiotics with antifungals, effectively alternating drugs when they can to avoid the emergence of resistance. Even with less life-threatening infections, we can work around the effects of antibiotics. Although there are many reasons we eat yogurt, those on antibiotics may benefit from the live cultures of *Lactobacillus acidophilus*. Or we might take capsules full of *Saccharomyces boulardii*, a kind of yeast that helps maintain the microbial balance. If we can keep our microbiome healthy, there is less chance of a hostile microbial takeover.

For much of our existence our microbiome has helped to keep potentially invasive microbes—particularly those already in residence—in check. Most of those microbes are bacteria that in both number and diversity have the fungi beat. There is a good reason for this: our body temperature. Many bacteria thrive at 37°C (98.6°F), the normal human internal temperature, but for many fungi our bodies are like Death Valley. Most fungi prefer temperatures between 12°C (53.6°F) to 30°C (86°F). We mammals simply run too hot. Like a healthy microbiome, our warmth protects us from fungal invasion. But now some scientists worry that our so-called temperature barrier is beginning to fail us.

•◉•

OVER A DECADE AGO ARTURO CASADEVALL, A MICROBIOLOGIST AND immunologist at Johns Hopkins Bloomberg School of Public Health, published a paper positing that fungi and their relative intolerance of warm bodies may have enabled the rise of mammals, a phenomenon he called the "Fungal Filter." "Mammals make no sense," he says. "We have to eat *so* much. Most humans eat four or five times a day. This is just not the amount of food consumed by most other animals on the planet. Most of that food is to maintain our temperature." We are high-energy animals, and Casadevall argues that something had to select for what would seem like an unfavorable metabolism. Until about sixty-six million years ago mammals were not the dominant species; for nearly two hundred million years before that, giant sauropods, stegosaurs, and theropods like *Tyrannosaurus rex* ruled Earth. Then an asteroid slammed into the planet. The impact caused the extinction of nearly 80 percent of all animal species, from the dinosaurs to marine invertebrates.

The post-asteroid period, according to Casadevall, "was associated with a massive fungal bloom, we know from the fossil records. Any surviving animals would have been exposed to fungal spores and potential pathogens." The limestone, shale, and sandstone formed from deposits laid down eons ago hold evidence of more than what was once bone. The fossil record written into these layers is a memoir of the worms, plants, insects, and microscopic specks of life from plant pollen to fungal spores: billions and billions of spores. After the cataclysm Earth would have been littered with dead and dying plants and animals—a feast for fungi. Earth didn't have a second reptilian age because fungus infected the more susceptible reptiles, while warm-blooded mammals survived the fungi. That's the hypothesis. And it is possibly why bacteria and viruses, which tend to tolerate the human hothouse, are typically more important pathogens compared with fungi.

Now, thinks Casadevall, climate change is challenging one of our important defenses against fungi. A warmer environment may enable some fungi to evolve a higher temperature tolerance. If a fungus can jump the temperature barrier, then humans and other mammals could become hosts to novel fungal infections. A yeast that normally grows in wetlands or on apple trees could possibly evolve to live in goats or bats or humans.

In 2010 Casadevall and Monica Garcia-Solache coauthored an opinion piece for a scientific journal. They hypothesized that warmer temperatures would alter and likely increase the geographic range of disease-causing fungi and likely select for *new* fungal pathogens with higher tolerances for a warm body. In 2019, less than a decade later,

Casadevall and colleagues suggested that the emergence of *C. auris* may be the first example of a climate-enabled, novel human fungal pathogen. The concern, he says, is that as we move into a century with higher temperatures, some fungi will adapt, breaking through the fungal filter. "Right now, most fungi in the environment are simply not able to survive our temperature, but if some adapt to higher temperatures, then we may be open to new pathogens. That's what I've argued with *C. auris*."

Casadevall imagines that a fungus that can normally grow in temperatures up to 36°C (96.8°F) when exposed to hot days could eventually evolve so that it survives at 37°C (98.6°F)—our range. It is possible that just one degree would be enough for selection. Fungi are known to adapt rapidly to temperature, which means that every really hot day, he says, becomes a roll of the dice. "We have argued this is the canary in the coal mine," says Casadevall, or in other words, a fungus all ready and loaded to go, infecting other species, maybe insects or reptiles, but that couldn't hack the warmth of a human body. "And then at some point—some adapted. Now we have a huge problem. There could be other explanations," concedes Casadevall, "but right now we don't have any."

When *C. auris* was first isolated from the ear of a Japanese patient in 2006, the sample had been archived by researchers who were simply isolating yeasts that could infect humans for an antifungal study. That same year the fungus was also found in a cluster of patients in South Korea with chronic ear infections. Then in 2009 it was found to have invaded the bloodstream of one elderly patient and two infants in South Korea. Of the three, only one of the infants survived. The "ear" fungus was now lethal. By 2016 the CDC viewed *C. auris* as an urgent threat, which is when it advised hospitals and other long-term care facilities in the United States to be on the lookout. By then the circulating strains seemed to have spread between patients and between hospitals. One hypothesis is that the fungus first infected the ear canal because it is naturally cooler than the rest of our body and so more tolerable—a first step into the hothouse.

• • •

IN 2019, BRENDAN JACKSON, A MEDICAL EPIDEMIOLOGIST AT THE CDC, along with other agency scientists, published a paper entitled "On the Origins of a Species: What Might Explain the Rise of *Candida auris?*" The team had identified four genetically distinct populations of *C. auris* that had emerged "nearly simultaneously" in East Asia, South Asia, Africa, and South America. The majority of cases in the United States likely hailed from the South Asian population. A subsequent paper by a different group added one other potential emergent population isolated from a patient in Iran. Scientists believe there are four "clades," or distinct populations originating from a common ancestor, and possibly more of *C. auris*. This astounding finding would be akin to having several different variants of the SARS-CoV-2 virus, which causes COVID-19, emerging in late 2019 all at once rather than originating from a single strain that evolved over time.

When reports of *C. auris* were first categorized as something "new," Jackson was skeptical. He thought maybe it was a pseudo-outbreak, a misdiagnosed yeast that had been around for years before disease detection technology improved. But data from an earlier publication that reported the absence of *C. auris* in more than ten thousand *Candida* samples collected between 1996 and 2009, quickly disabused him of that notion. A subsequent review of samples collected between 2004 and 2015 identified only four cases and only one prior to 2013. The fungus is more remarkable for the separate clades that are so geographically dispersed. Where did they come from? And why now?

To understand how weird the pattern of emergence for *C. auris* is, consider that within months of the occurrence of SARS-CoV-2, it was traced to a single region in China and discovered that at some point in time the virus may have jumped from bats or pangolins or some other wild animal to humans. The Ebola virus too emerged in a single area in Central Africa before radiating outward to other regions. A recently emerged fungal disease called *Sporothrix brasiliensis*, which is transmitted from cats to humans (and from cat to cat), has turned up in Brazil, Argentina, Paraguay, and Panama. First identified as a clinical oddity in Rio de Janeiro in 1998, it moved across the Americas carried by its feline hosts. It is not unusual for a disease to emerge in one place and travel the world (mutating as it goes), but for a disease-causing microbe to pop up in different geographic areas with different genetic signatures at the same time is odd.

In 2019 the novel invader, *C. auris*, got an inadvertent boost: COVID-19. As the SARS-CoV-2 virus sent millions to the hospital, *C. auris* was there lying in wait. All the fungus needed was an influx of morbidly ill bodies and a distracted health system to continue its rise. One Florida hospital experienced a fungal outbreak during a 2020 viral surge. Of the fifteen *C. auris* cases that had been detected, twelve were in COVID-19 patients. The outbreak was found to have originated from a single source, perhaps a newly admitted, critically ill patient. When the virus surge subsided, so too did the fungal outbreak. As the tide turns and the viral pandemic recedes and becomes another endemic human virus, *C. auris* will not only remain a threat but could become more prevalent.

•••

C. auris IS A TRIPLE THREAT: DRUG RESISTANT, RESILIENT, AND HERE to stay. And there will be others. Fortunately, most of us are not completely defenseless. If we have one superpower against pathogens, it is our immune system. Every day we are exposed to thousands of different microscopic organisms, including dozens of yeasts and other fungi. Skin is our first line of defense, setting up a physical barrier protected by a secondary web of bacteria, fungi, and other microbes in our bodies. Our lungs are like an intricate, branching tree of life with just a single cell standing between what is inside our bodies and outside. Arguably our most vulnerable organs, lungs are protected by an ever-moving biological escalator of mucus moved along by an underlying layer of cells coated with a mop of hair-like protrusions called cilia. The mucus acts like a lint roller trapping microscopic debris—including the steady supply of spores, dust, and pollen—and the ciliated cells move the whole mess along, away from our lungs. When we sneeze or cough, we expel that debris-laden mucus. Some diseases like cystic fibrosis prevent this system from working well, leaving those with the disease more susceptible to infection for lack of the often-unheralded physical barrier. Our digestive tract is also lined with mucus-producing cells that help protect us from infiltrators.

Sometimes these frontline systems fail, which is when our cellular immune response, the macrophages, T-cells, and others kick in. These are the cells that gravitate toward and ingest or kill invaders; some also release an arsenal of chemical defenses. The outcome may be fever, inflammation, or other collateral damage, causing a body to feel ill, a side effect of a robust defense. Fever protects against invaders that can't tolerate the heat. This nonspecific response is buying time for a more powerful and targeted defense, the adaptive response. Cells called T-cells are a key member of this system. These are the cells that kill infected cells, recruit others, and help regulate the immune response. Another type of immune cell called B-cells make exquisitely specific antibodies targeting the pathogen. These are the cells that also provide immune memory—the ability to fend off the same bacteria or virus or fungi a second or third time around. When we are vaccinated, we are provoking this response so it will be there for us for any future exposures. This system has protected vertebrates from frogs to humans in one form or another for hundreds of millions of years. It isn't perfect, but it is so multipronged that if one strategy fails, another may do the trick.

"If the T-cells don't kill an invader, your neutrophils will," Stuart

Levitz tells me. "Or macrophages. For some organisms, only when multiple systems fail, do you get sick." Levitz is an infectious disease physician and mycologist at the University of Massachusetts Medical School, whose work centers on how immune cells respond to fungi. He has a favorite Gary Larson cartoon he likes to show medical students in which firefighters hold a net for a woman jumping from a burning building. She rebounds after jumping into the net and goes through a window into another burning building. "For a lot of organisms, the immune system is like that": survive one thing only to be thrown into another. So in this regard *C. auris* is no different than many other fungal pathogens that have a hard time infiltrating bodies with a healthy immune system.

An invasive fungal infection used to be a rare occurrence in the clinic. "If you had a fungal infection," recalls Levitz of his early days in medicine in the 1980s, "the case would be presented at a conference," that's how rare they were. "Now we see them all the time." The reason is because we live in the age of the immunocompromised. A growing global population lives with immune systems that are compromised to some degree. Advances in organ transplantation have enabled countless people young and old to live full lives with transplanted kidneys, lungs, hearts, and other organs. In the United States nearly forty thousand organ transplants are done each year. All of these patients depend on immune-suppressing drugs to reduce the chances of rejection—and some require these drugs for the rest of their lives. Cancer survivors, depending on the type of cancer, may also be living with compromised immune systems, as are many of us who are simply aging or use powerful steroid medication to control conditions like asthma, chronic obstructive pulmonary disease (COPD), and cystic fibrosis. The trade-off of these fantastic feats of medicine is in some cases a compromised immune system. We are living longer and better but are increasingly becoming more susceptible to invasive fungi.

Levitz began his practice when the human immunodeficiency

virus, or HIV, which causes AIDS, was on the rise. The virus targets CD4 cells, or T-cells. AIDS hadn't yet peaked in the United States when Levitz began practicing, but he says of those earlier days, "We were seeing all these cases in the hospital of a fungal infection called *Cryptococcus*," a fungus that seldom invades immunocompetent hosts. It was as if HIV was in cahoots with bacteria and fungi that couldn't normally make a living on humans, but by suppressing our immune system the virus opened the gates and let the opportunists in. Levitz has continued studying and treating patients with *Cryptococcus* and other invasive opportunistic fungal infections.

Four decades later, for those able to receive treatment, HIV is no longer the death sentence it once was. Still the numbers of people infected are staggering. Around the globe more than 28 million people infected with HIV are living with antiviral therapy. This accounts for about 75 percent of all people with the virus, which means there are 9.7 million who are not receiving therapy, with millions more diagnosed every year. Hundreds of thousands are likely to die from cryptococcal meningitis each year. The AIDS virus, along with the rise of an immunocompromised population and heavy use of antibiotic drugs, has provided fertile ground for opportunistic infections, and when our natural defenses fail or are unable to respond, antifungal medications are the only recourse.

Antifungal drugs are chemical warfare waged on a microscopic field. The trick is to destroy the enemy while leaving the civilians unharmed, which can be difficult if they are hard to distinguish from one another. Despite the similarities between yeasts and bacteria, fungi are more animal than they are bacterial. All fungal cells, like our own cells, are eukaryotic. Most prominently, our cells are distinguished from those of others, like bacterial cells, because our genetic material resides in the nucleus. Our cells and those of fungi also share other structures and have similar composition, which in turn makes fungal cells difficult to target and kill without damaging our own cells. In contrast, antibiotics like penicillin are relatively safe because the drug targets a component of the bacterial cell wall that we do not have. The powerful antifungal drug amphotericin B, introduced to the market in 1959, was a lifesaver. It works by creating hole-like pores in the membranes of fungal cells, interfering with the normal function of fungi. One target for antifungal drugs like amphotericin is a chemical called ergosterol that fungi need to make the membranes surrounding their cells. Our cell membranes don't have ergosterol but do rely on cholesterol. The two molecules share enough basic chemical structures that a chemical that targets one may inadvertently target the other, causing potentially deadly side effects like kidney failure. Levitz says physicians used to refer to it as amphoterrible. Less toxic formulations are available now as are other less toxic classes of antifungals. But there are not many.

Drugs are classified by how they kill or by which part of the cell or cellular mechanism they target. There are three major classes of antifungal drugs used to treat systemic infections: polyenes (which include amphotericin and nystatin), azoles, and echinocandins. For comparison, there are over a dozen classes of antibiotics that target bacteria. Amphotericin targets ergosterol. Azoles target an enzyme that fungi rely on to make ergosterol; we have a similar enzyme that may be affected by some azole drugs. The echinocandins prevent fungi from making another essential chemical used in building the fungal cell wall called 1,3-β-d-glucan. Our cells don't use this molecule—which helps reduce the risk of side effects.

Antibiotics and antifungals are both powerful types of drug. And yet both are losing their efficacy as their targets, bacteria and fungi, evolve resistance. When we are infected and colonized by disease-causing microbes, they reproduce and sometimes we get sick. Anytime a cell reproduces, including our own cells, DNA is replicated and then divvied up into new daughter cells. This happens whether cells make clones or reproduce sexually. Whenever DNA reproduces, mistakes, or mutations, are inevitable. Some are corrected. Others are enough to kill the cell. Many may not matter one way or the other. A few may be helpful, like a change in an existing enzyme that allows a cell to detoxify and survive exposure to a lethal chemical, including antibiotics and antifungals (for the targeted microbes, these are lethal chemicals); or a mutation that allows a cell to pump out harmful chemicals; or a tweak in a protein so that it hides a key molecule targeted by a drug. New mutations are just one way in which microbes evolve resistance to the drugs we use. A microbe may also acquire resistance from other microbes; bacteria are notorious for sharing resistance genes. Scientists are finding that in some cases genes conferring resistance to some of today's drugs (particularly those derived from nature, including many of our antibiotics), have existed for millennia—having evolved in bacteria well before we began pressuring them with antibiotics. Survival doesn't always require resistance genes. Some microbes and fungi including *C. auris* may also survive exposure to drugs and other environmental challenges by forming biofilms, a microbial collective in which bacterial or fungal cells on the outside are sacrificed to protect those on the inside. The plaque in our mouth is a biofilm, as is pond scum. Biofilms are a hallmark of resistance because they are difficult for drugs and other chemicals like disinfectants to penetrate.

Fungi that are resistant to more than one class of drug are called multidrug resistant. Those that resist all drugs are pan-resistant. They cause an essentially incurable infection. In the summer of 2021 the CDC reported two separate outbreaks of pan-resistant *C. auris*, which resisted all three major classes of antifungal drugs. What is oddest about the yeasts' ability to thwart our antifungal drugs is that unlike other fungi and bacteria that have had decades of exposure to antimicrobials, *C. auris* is a novel human pathogen, and so no one can explain how it came by its remarkable drug resistance. But even pathogens that resist a single drug can be problematic.

Some 90 percent of *C. auris* isolates (yeast cultured and tested) in the United States are resistant to fluconazole, a popular antifungal drug. This is a problem because sometimes patients are treated with fluconazole to prevent *other* Candida infections of the mouth, urinary tract, gut, vagina, or elsewhere "and then that can lead to them being colonized by *C. auris*," Levitz says. So in some cases using the drug inadvertently paves the way for colonization and infection. "Everything is in competition in the microbial world," he reminds me. In some cases the top competitor may be a potentially lethal infection.

• • •

WHEN A NEW PATHOGEN LIKE *C. auris* EMERGES, ONE WAY IT TRANSITIONS from wherever it has been living into a new host may be the acquisition of new traits. Imagine that in its hourly struggle to survive over long periods of time and many generations, a fungus species might acquire a protective capsule—a bit of coating—that shields it or even disguises it from other microbes or cells. Then it acquires some enzymes that enable it to survive whatever chemicals other microbes might lob at it, or perhaps it is even exposed to antifungal chemicals in the form of agricultural pesticides. If it can overcome these chemicals, it may also overcome the same or similar chemicals used as antifungal drugs. Maybe it also evolves to tolerate warmer temperatures. Now we've got a yeast that once made its home in an apple tree or in a wetland but that at this point can live quite happily in our body, evade our immune system, and disarm our drugs. Then some of us carry it from one country to another and then another, and eventually it finds a host in a hospital patient who has recently received an organ transplant or is elderly with a weakened immune system.

The deadly yeast isn't the only fungal infection on the rise. In summer 2018 students returning from a service trip to Tijuana, Mexico, arrived in New York City with an odd pneumonia. Physicians treated them with antibiotics, which failed to cure them, and the CDC was called in. Epidemiologists found that other volunteers who had participated in similar trips around Central and North America had also fallen ill. All were eventually diagnosed and treated for a *Coccidioides* infection, or Valley fever.

The students prior to infection had been healthy with good immune systems. Unlike fungi that favor hosts with compromised immune systems, *Coccidioides* has no such limitation. The fungus is endemic in much of the southwestern United States and parts of Mexico and Central America where the climate is warm and dry. Just knowing the location where a patient contracts the disease is helpful for diagnosis, particularly for microbes known to be prevalent in certain regions and absent from others. Where particular fungal infections are rare, physicians aren't always thinking "fungus," let alone a fungus that is endemic to some other region of the country, as with the cases of Valley fever in New York City. In 2019, 20,003 cases of Valley fever were reported, mostly in residents of California or Arizona. On average 200 Americans a year die from it. Scientists think the fungus will likely spread further into the United States as the climate changes; one model suggests that by the end of the century the fungus will affect much of the West all the way up through Montana.

In 1999 a fungus called *Cryptococcus gattii* was found to be infecting humans and animals who lived on or had visited Vancouver Island in British Columbia. Porpoises, dolphins, and other marine mammals died from the fungus, which typically lives in soil and on several native trees. When its spores are released into the air or settle on coastal waters, they can be inhaled, infecting the lungs, brain, and muscle. The lethality rate is high. Before 1999 *Cryptococcus* had only been known as a problem in tropical or subtropical locales. Its appearance in the Pacific Northwest was unusual. Arturo Casadevall and a colleague hypothesize that the fungus likely arrived by ship earlier in the twentieth century, possibly in the ballast water carried by ships to increase stability and that is released when extra weight is no longer needed. Then in 1964 a tsunami that flooded the coastal forests may have carried the fungus to shore. Once the fungus colonized the region, human activity like land-clearing and logging may have aided its dispersal into the air, where spores that landed on surface waters infected marine mammals.

In India, during the 2021 wave of the COVID-19 pandemic, a fungal infection called mucormycosis, caused by a group of related fungal molds, killed thousands. Hospitals that had seen only a couple of cases of the fungus a year were suddenly identifying hundreds of cases. Before the pandemic the fungus killed 50 percent of those infected. During the pandemic the mortality rate rose to 85 percent. The fungi that cause mucormycosis had been known to infect humans, particularly those who had diabetes or were immunocompromised, but never to the extent that it ran rampant during the viral pandemic. The eyes, nose, skin, bone, lungs, and other regions can all be infected, and infection can quickly lead to death. Treatment often requires antifungals combined with aggressive surgery to excise the fungus and damaged tissues, leaving many survivors disfigured. As with *C. auris*, the pandemic provided an opportunity for the fungi to reach vulnerable patients. Normally found in soil, compost, animal dung, and other environmental sources, the fungi that cause the disease are ubiquitous. Although most cases occur in India, mucormycosis infections have been reported around the globe, including in the United States. Though each deadly and disfiguring fungal infection in humans is tragic, humanity has yet to experience a full-on fungal pandemic. The same cannot be said for other species.

Chapter 2

EXTINCTION

Karen Lips studies amphibians in Costa Rica and Panama and across the Americas. She is best known for her work with frogs and, inadvertently, for documenting their decline. When Lips began graduate work in the late 1980s, she imagined adventure and faraway places. Maybe she would study lizards in Australia or frogs in Central America. Her PhD adviser, Jay Savage, was well known in the herpetology world and was at the time writing what would become the definitive book on the ecology and evolution of the Costa Rican local fauna: *The Amphibians and Reptiles of Costa Rica*. The first summer in the field in Costa Rica, Lips found a site in the middle of nowhere inhabited by a spectacular thumb-sized frog. It was "fluorescent green, spiny, and just incredible," she recalls. "It blended in with moss on trees." But down the mountain the same species was brown and smooth. The highland frogs also had different songs from the lowlanders. The frog's scientific name had been *Hyla lancasteri*, and Lips—who subsequently renamed the highland frogs *Isthmohyla calypsa*—could easily imagine how the frogs' oddities could provide a PhD's worth of experiments.

She returned to the field site in Costa Rica's rugged Talamanca highlands and lived in a shack owned by a local family. In the lower river valley the family had cleared land for pasture, but at the higher elevations and surrounding the farm, the cloud forest was old growth, undisturbed and filled with incredible species. It was a forty-five-minute hike from the family's house, and the house was an hour's drive down the mountain, often requiring four-wheel drive and chains on the tires, to the nearest town with bus service. There was no running water, phone, or electricity, but the mossy, lush, drippy cloud forest was a herpetologist's heaven. She began her PhD field work in the spring of 1991 and remained on site for the next two years. She spent her days walking on moss-covered rocks strewn through streams, measuring tadpoles, and looking for egg masses on plant leaves. At night she sought out nocturnal frogs and other creatures. In December 1992 Lips traveled home for the Christmas holiday. When she returned in the new year, there were fewer frogs. By the following year they were all but gone. Everything else was the same. The site was in a protected area, La Amistad Biosphere Reserve, and there had been no trees cut or roadways built. The only difference seemed to be the lack of frogs. The few frogs she did locate were dead or dying. It's rare to see dying or even dead frogs, she says, since snakes and birds usually make quick work of the weak or freshly dead. She sent seven of the moribund frogs she discovered along with a few healthy frogs for comparison to a pathologist who noticed something odd in their skin but didn't know what it was and couldn't identify the cause of death. Absent any obvious reason, Lips wondered if the frog decline at her site might be related to her own techniques, "researcher disturbance," or toxic chemicals or altered rainfall patterns or disease—perhaps some bacterial or viral epidemic. There were a lot of questions and no answers. After defending her dissertation but deprived of frogs to study in Costa Rica, Lips relocated to a new study site in western Panama.

In the late 1980s scientists were aware of and concerned about vanishing frog populations in the United States, Australia, Costa

Rica, and Mexico. Forests and tropical villages once raucous with frog song went quiet. There were all sorts of theories: climate change, habitat loss, drought, ultraviolet light, pesticides, and virus. Some losses were so large and so sudden that they were controversial. Were they really "losses" or simply the ups and downs of natural populations—ecology in action? Before Lips began working in the cloud forest, golden toad populations, which were once numerous in Costa Rica's more northern Monteverde Preserve, crashed. One observer said that for well over a decade before the collapse, depending on the year, hundreds or thousands of the toads could be found. There was no obvious explanation for the decline, but whatever was happening was not in any way usual.

Frogs begin their lives as tadpoles in a watery nursery. Fish-like frog larvae breathe through gills that extract oxygen from water as it flows over highly vascularized rows of respiratory cells. Like other amphibians, frogs eventually develop lungs that allow them to live on land, though most won't stray very far from water and almost all will eventually return when it is time to breed. Some spend their lives in water. A few, including many of the colorful poison dart frogs, do things in reverse. The strawberry poison dart frog breeds on the forest floor and then carries each newly hatched tadpole on her back, releasing it into a tiny pool of water that has collected between layers of bromeliad leaves—one tadpole per pool. She also drops unfertilized eggs into the pool so that the tadpoles have some sustenance. The gastric brooding frog, when it existed, also reproduced on land. To provide a suitable aquatic nursery, a brooding frog mother would swallow her fertilized eggs to mature her froglets in her stomach. The species went extinct in 1983; no one knows exactly why, but the culprits range from habitat destruction to pollution and disease.

Frogs and other amphibians and reptiles are ectotherms, animals with little internal ability to control their own temperature. If the weather is warm, they are warm. If it is cool, their bodies are cool.

A type of wood frog that lives in Alaska virtually freezes for seven or eight months of the year; its body hardens slowly while its liver begins pumping out sugars that act like antifreeze, ensuring survival. Eventually ice crystals grow in its gut and in spaces around the heart, brain, and other organs. The frog eventually freezes solid and then revives when the warmer temperatures return. But the myth about frogs in boiling water not noticing the heat until it is too late isn't true. Given the opportunity to hop out, they will.

Frogs vary not only in color but in size too. Some are thumbnail-sized, and others are large enough to eat birds and mice. Almost all have teeth, drink through their skin, and have sticky tongues that they use to catch their prey. Most frogs are carnivores with varied diets: larvae, flying insects, snails, and whatever else they can catch on their tongue and stuff into their mouths. (Tongueless frogs use their front legs to snatch food.) The frog's predators are many, including birds, small mammals, and snakes, and some frogs defend themselves by secreting toxic chemicals through their skin. Poison dart frogs, *Dendrobates*, are armed with bitter, highly toxic alkaloids. Their colorful skin warns away would-be predators. Most scientists agree that the frogs don't make their poison but acquire it from their diet. Alkaloids are complex, often biologically active chemicals, which makes them potentially useful to researchers seeking new cures for pain, cancer, and other ailments. Some alkaloids are hallucinogens.

Frog skin is so much more than color and texture; it is a remarkable organ. It is permeable to oxygen, which helps frogs breathe. It is also studded with mucous glands that secrete slime to keep the skin moist and is loaded, like ours, with a complex and potentially beneficial microbiome. Skin is a first line of protection against disease, and frogs, like other vertebrates, also have a relatively advanced immune system with differentiated immune cells like T-cells and macrophages and antibody-making cells. Frog skin also regulates electrolytes like salts, sugars, potassium, and other essential molecules. This reliance on the functionality of skin makes frogs particularly vulnerable to skin diseases.

. ● .

ANIMALS, INCLUDING HUMANS, AND FUNGI SHARE A COMMON ancestor: a single cell with a beating tail called a flagellum, which propelled the cell forward through the watery environment much like a sperm cell. Then, sometime roughly one billion years ago (more or less), the ancestors of animals and fungi parted ways. Their descendants evolved and diverged, blossoming into the complex, interconnected tree of life we know today. In time, plants, animals, and fungi, all once wholly aquatic, moved onto land. Plants set down roots; animals crawled, swam, and flew; and fungi intermingled with all things plant, animal, and bacteria. Most fungi lost their ability to move on their own and instead relied on air currents and traveling animals to carry spores. A type of fungus categorized as Chytridiomycota, or chytrid fungus, diverged early from the fungal family tree and kept its tail-like appendage. Chytrid spores are mobile, and like many other fungi, they are saprobes, mostly feeding on the dead and eating from the outside in—digesting its food and absorbing what remains. Some also parasitize the living. When the fungus settles onto a food source, long, probing, root-like structures called rhizoids burrow into host tissues. As they feed, chytrids, like other fungi, secrete enzymes, dissolving plant, animal, or microbe into food. A full-fledged chytrid, or thallus, looks something like a sinister smiley-face button that has grown an extensive set of roots. Over time a rounded structure, a zoosporangium (a fungal vesicle or sporangium) that fills with the flagellated zoospores, takes shape. Once the spores are released, they seek out a new host. Because the zoospores are motile, they have some agency. Some chytrids target algae, while others infect fungi and other hosts. The fungus seeks both chitin and keratin in its food. As it turns out, amphibian skin is a good source of keratin.

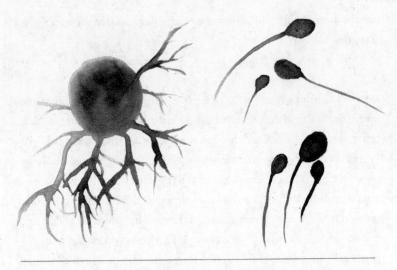

CHYTRID THALLUS / CHYTRID ZOOSPORES

When Lips moved her field site to Panama, she was hopeful. Here was a new project in a new location hundreds of mile away. She had set up transects (linear paths) in the forest to study populations of frogs, snakes, salamanders, and lizards. "We'd be figuring what's there and ask cool questions." It was a place she could return to and study for decades. But in December 1996 it became clear something was wrong. The frogs were easily spotted during the day or were sitting on the ground rather than on the leaves that hovered over the streams. "We would grab a frog, and it dies in your hand." At some point she recalled wondering if this was one of those mysterious frog decline sites. "It was like you are in the middle of a mystery but don't know it."

Between 1996 and 1998 frogs at the National Zoo in Washington, DC, including blue poison dart frogs, green and black tree frogs, and White's tree frogs, died from an unknown skin disease. Scientists suspected chytrid fungi but couldn't identify any particular species, so they sent samples to Joyce Longcore, a mycologist affiliated with the University of Maine who was well known for wrangling and identifying the odd fungi. Longcore described the sample as a previously unidentified chytrid species. A year later, in 1999, she and the zoo's scientists gave it a name when they published a paper describing *Batrachochytrium dendrobatidis*, or Bd, a chytrid pathogenic to frogs and one of the first known to parasitize a living vertebrate. Instead of exploring the quirks of an unusual frog, Karen Lips had documented the rise of Bd—a frog-killing chytrid fungus.

When the Bd chytrid settles onto a frog's skin and germinates, it sends out its rootlike rhizoids. There are some indications that the invading fungus can even suppress the frog's immune response. Once inside, the chytrid feeds on keratin and other nutrients, cloning itself until its zoosporangium is full. When a frog is infected, hundreds or thousands of these sporangia can form. As they each fill with zoospores, the zoosporangia rise on plant-like stalks through the frog's skin. Millions of motile zoospores will be released into the environment in search of new hosts. The disease leaves the frog's skin in tatters. Infected animals struggle for oxygen, and electrolytes are thrown out of balance. They die through a heart attack–like event.

It has been over thirty years since Karen Lips first traveled to the Costa Rican cloud forests to study frogs. In 2019 dozens of scientists, including Lips, wrote of the destruction caused by Bd: "This represents the greatest documented loss of biodiversity attributable to a pathogen." Before Bd, no one knew any disease could be *so* bad. Now they know.

• • •

THE AFRICAN CLAWED FROG (*Xenopus laevis*) ISN'T PRETTY. IT COULD even be called homely, with its broad, flat head; warty, mud-brown body; and long, slender fingers. Unlike many other frogs, it lacks a tongue and can live its entire life in a tank of water. And yet the sub-Saharan frog is one of the most highly traded amphibians in the world.

The frog's travels began shortly after 1928 when Lancelot Hogben, an English endocrinologist, landed in Cape Town, South Africa. Hogben was investigating the role of the pituitary hormones on frog skin. The clawed frog changed color depending on the environmental conditions. It could be so dark it was nearly black or so pale as to appear almost white. When Hogben removed the pituitary—a small gland that rests just under the frog's brain—the frogs remained almost white. Determined to confirm the role of the little gland, Hogben injected a female frog with pituitary hormones from an ox as a sort of hormonal replacement. She in turn released eggs. It didn't matter if a male was around to encourage egg release. Ox hormone induced the frog to ovulate.

Hogben and colleagues published their findings on the pituitary and ovarian function in African clawed frogs in 1931. A short time later, scientists found that the human gonadotropin hormone hCG could do the same: it also caused frogs to ovulate. The hormone circulates in the blood and is excreted into the urine of pregnant women. It wasn't long after that discovery that the frogs were put to use in pregnancy tests, eventually known as the Hogben test.

The frogs were a revolution. Pregnancy testing at the time relied on rabbits—the proverbial "rabbit test." It was a messy, lethal affair involving a syringe full of urine and eventual death for the rabbit, which was killed and dissected days after injection. An ovulating rabbit meant a pregnant patient. The old saying "the rabbit died" doesn't necessarily mean a positive test because *all* the rabbits were killed. The frogs promised a less fraught test. Physicians simply injected them with a woman's urine sample, and the frogs would either release their eggs into the aquaria in which they were held or not. Ovulating frogs yielded an answer within a day and put an end to the killing and dissection of rabbits. Because the frogs lived in a tank of water, they were easy to keep in a lab or ship around the world. Throughout the 1930s and into the late 1960s many thousands of African clawed frogs were ovulating for women all over the world. By 1970 clawed frogs, kept either as pets or in laboratories, were among the world's most widely distributed amphibians.

Many of the frogs used for testing had been shipped from the Jonkershoek Inland Fish Hatchery near Stellenbosch, South Africa. Because the hatchery was unable to successfully breed the frogs, collectors were paid to gather animals from the wild for about sixty-five cents apiece. By 1969 the hatchery shipped more than four hundred thousand frogs; half of those shipments were international. By then, reports began to emerge of African clawed frogs living in lakes and ponds on the Isle of Wight and in Arizona, Florida, California, and elsewhere in the United States, Europe, and Japan. Populations were found in Chile and Portugal too. Combined with frogs shipped by private suppliers, clawed frogs were swimming around in tanks on every continent. Some inevitably escaped, while others were "humanely" released into local ponds after they had outgrown their welcome as pets or their usefulness as test subjects.

Much of Bd's spread happened before anyone knew what it was. Without going back in time, it's hard to re-create how the chytrid traveled and what populations it hit along the way. The fungus moved with the frog trade, but which species carried it where remains a mystery.

Vance Vredenburg, a disease ecologist at San Francisco State University, tracks diseases through time and space using preserved museum samples. For centuries collectors enthralled with the diversity of nature have scoured forests, streams, and mountains for frogs, snakes, bats, and other creatures. If they weren't bringing an animal back home alive, they gutted and dried it or stuffed it in a jar of formalin. The latter would then be stored in ethanol in perpetuity. These pickled samples have a shelf life of hundreds of years. But the initial formalin shock jumbles the DNA. In the early days collectors didn't know about DNA (the composition of genetic material was largely a mystery until the mid-twentieth century). And never mind any one of them imagining that scientists of the future would be combing through their samples to unlock the mysteries of a pandemic fungus.

MOUNTAIN YELLOW-LEGGED FROG

Between 2004 and 2008 Vredenburg studied the mountain yellow-legged frog across dozens of ponds in Sequoia and Kings Canyon National Parks in California. Over the four-year period he witnessed a landscape covered with thousands upon thousands of dead frogs. The devastation transformed Vredenburg's view of science and of nature. "We were so wrong. We had a hundred years of disease ecology, and we were wrong about just how destructive a single pathogen could be." He now keeps jars filled with hundreds of dead frogs in his office to remind him of how quickly things can change.

Vredenburg wondered if there might be a link between the globe-trotting African clawed frogs and the chytrid pandemic. The African clawed frogs carry Bd chytrid but don't get sick, which makes them asymptomatic carriers or even potential super-spreaders. To test his hypothesis, Vredenburg turned to DNA. He sampled frogs (dead or alive), amplified the DNA, and then looked for a particular tell-tale fungal sequence. If the sequence was there, in most cases the chytrid fungus was there. The method he used, developed by a lab in Australia, required a mere snippet of DNA, which enabled him to study the archived frogs with the jumbled DNA. The genetic marker he sought was relatively small and abundant. The test, he says, is somewhat similar to the COVID-19 swab test. The downside is that it is a relatively coarse test for the fungus generally. He could identify that the fungus was present, but distinguishing different lineages of the same fungus would require deeper genetic sequencing.

In 2012 Vredenburg fished some frog samples from several old museum specimens that had been collected and preserved from Kenya and Uganda in the 1930s or before. Three of the samples collected in Africa during the 1930s were chytrid positive. The timing, the spread, and the clawed frog's lack of sensitivity to the fungus were all suspicious. There is a disease dogma that suggests over time a disease becomes less pathogenic for its own good. A pathogen that kills fast and often will die out with its host. Usually, the longer a host population has lived with a disease, the more likely it is to become either resistant or tolerant. The data suggested to Vredenburg that one way the fungus moved was by hitching a ride on the backs of the African clawed frogs. The story of how Bd spread around the globe is incredibly complicated, and there are likely multiple routes and multiple species with which the fungus traveled, he says. "But it looks like the pathogen's *origin* is Asia."

The study that nailed down the chytrid's origin was led by Simon O'Hanlon and his postdoctoral adviser Matthew Fisher and included Lips and dozens of other scientists. Fisher, a professor at Imperial College in London, studies emerging pathogenic fungi across species. Before this study there were several places around the world from which chytrid was thought to have originated, including North America, Japan, South America, and East Asia. Another unknown solved by the study was the age of the lineage that swept the world, a highly virulent lineage labeled appropriately as Bd-GPL, for global panzootic lineage. One study suggested it emerged about a century ago, and another suggested emergence occurred twenty thousand years ago. When a population of fungi (or any organism) separates from others in time or space or both and individuals reproduce, small, random genetic changes can occur. Given enough time and enough changes, populations can be identified as different lineages. Sometimes the changes allow disease trackers to discover the age of a lineage or the route it traveled. O'Hanlon and colleagues sequenced the genomes of hundreds of chytrid samples collected from all impacted continents. Their results pointed to a common ancestor located somewhere on the Korean peninsula between 50 and 120 years ago.

That there are several species of frogs in the region that carry the fungus but aren't harmed is another clue to its origin. The finding of asymptomatic frogs with Bd chytrid also suggests that animals collected and traded from locales around the Korean peninsula are possible carriers. From its point of origin the fungus then traveled the world and evolved into several different genetic lineages. O'Hanlon and colleagues' conclusion is sobering: "Ultimately, our work confirms that panzootics of emerging fungal diseases in amphibians are caused by ancient patterns . . . being redrawn as largely unrestricted global trade moves pathogens into new regions, infecting new hosts and igniting disease outbreaks."

• ▪ •

AT ANY POINT IN TIME MILLIONS OF ANIMALS—IN CAGES, CRATES, and tanks—are moving around the world. In the United States alone, two hundred million animals arrive at the ports every year, which translates to roughly six hundred thousand individual animals a day, plus three tons more because some animals are simply counted by weight. Over two thousand *different species* from across the animal kingdom, including hundreds of species of birds, reptiles, amphibians, monkeys, and spiders hailing from Singapore, Hong Kong, Peru, and elsewhere, land on our shores with little fanfare. Some are bought and sold by food markets and restaurants. Others are sent to research laboratories. About half of all live imports are destined for pet shops and eventually living-room fish tanks or basement terraria. In a ten-year period the European Union imported more than twenty million reptiles alone. This legal trade in animals fuels what some scientists call one of the "largest and most complex commerce exchanges in the world." Others see it as one huge conveyor belt of disease.

Many of the millions of animals moving around the world are traveling aboard massive container ships. The largest container ship carries the equivalent of more than twenty thousand shipping containers. Add air travel, and the number of animals moving around the world at any moment, legally and illegally, is extraordinary. For six years herpetologist Jonathan Kolby worked as a US Fish and Wildlife Service inspector in Newark, New Jersey. The port is the third busiest of the dozens of ports in the nation through which animals once wild, now crated and caged, including frogs and other amphibians, make a final stop before dispersing to pet shops and elsewhere. Kolby, now a conservation biologist and wildlife trade consultant, wrote an article about his experience as an inspector that was published in *National Geographic* in 2020. In it he describes the millions of animals arriving in the United States as a "kaleidoscope" of life moving about where any new or potential pathogen can easily slip by.

If those frogs and birds or fish and snakes never set foot or fin outside of the home or shop, trade might be of little environmental consequence. But with these kinds of numbers, escape or release is inevitable. Burmese pythons stalk the Everglades, consuming local fauna at an alarming rate: the primary source of these exotic snakes are one-time pet owners who released a pet grown too large or otherwise unwanted into the wild. A few lionfish—spiny, poisonous, aggressive, and prolific spawners—were released into Florida's warm waters, likely tipped out of aquaria. With no known predators in their new environment, the lionfish population exploded. These voracious feeders consume dozens of local species. One biologist commented that she had never seen a fish "colonize so quickly over such a vast geographic range." Snakehead fish, native to parts of Africa and Asia, are invading US waterways and ponds. They are believed to have been released by hobbyists or by those intending to grow the fish for food. Some may have escaped confinement through the practice of "prayer animal release," a faith-based practice of releasing captive animals into the wild as an act of compassion, but the outcome may be "at odds" with the intent.

Matt Armes is one of the many aficionados of exotic reptiles. Growing up in Norfolk, United Kingdom, he kept lizards in his bedroom and learned about nature from television personalities Steve Irwin and David Attenborough. As a teenager he worked in an exotic pet shop selling amphibians, reptiles, fish, and invertebrates. It was a thrill, he says. Instead of a few species kept in his home, Armes began breeding and keeping hundreds of different animals and dozens of species for the shop to be sold to other gecko, lizard, and snake devotees like himself. Working in the animal trade, his herp world exploded. The electric blue day gecko (*Lygodactylus williamsi*) in particular, he says, is stunning. Females are green-turquoise, and males in full breeding color are a dazzling blue. The gecko is native to a small region of Tanzania, is critically endangered, and is illegal to sell as wild caught.

Armes is now an ecologist spending his days capturing and monitoring reptiles and amphibians in the United Kingdom. He no longer works at the shop and intends someday to return to graduate school; maybe he will even help save the electric blue gecko. But he worries, based on his past experiences, about the potential for traded animals to introduce new diseases capable of infecting native animals. Most exotic pet owners, he says, are passionate about the animals they keep and don't intend harm. But the traders, hobbyists, and others who come in contact with the animals are not always well educated on how to keep, breed, and trade disease-free animals. It isn't all their fault, he says. Many just don't know any better, and globally, for most traded animals, there are no requirements for disease testing or monitoring. The result is an animal trade that has become a free-for-all for fungal pathogens.

The epidemiologist Matthew Fisher puts it this way: "Everything you move out of one environment across a continent is colonized by something that you probably don't understand and is potentially a threat." Every imported species brings with it its own community of viruses, bacteria, and fungi. Given the opportunity, a minor pathogen to one species can spiral out of control and go pandemic in another. Compared with a generation or even two generations ago, when the African clawed frogs and others began making their rounds, the rate at which a disease can travel the world today is shocking.

There are national and international laws for endangered species and for species known to be a risk for becoming invasive themselves. But these are a fraction of the animals moving around the world at any point in time. Most frogs and fish and other animals that pass through ports in New York or Los Angeles destined for pet shops are not examined for disease, and there is no law requiring that anyone does so. And these are the *legally* traded animals. There is a huge illegal trade in animals. Armes mentions the green tree python, a vibrantly colored snake native to Indonesia, which is illegal to capture and sell. Farm-bred snakes can be exported legally from their country of origin. But by the numbers more snakes are sold than could possibly be accounted for—which means the majority of snakes sold are wild caught and laundered through the pet trade as captive bred. The very nature of the illegal animal trade makes its magnitude difficult to quantify, but recent studies suggest that the numbers and impacts of animals (and plants) moving around the world illegally is grossly underestimated.

The global animal trade has created what some scientists say is "a functional Pangea for infectious diseases in wildlife." *This* is what worries scientists like Armes, Vredenburg, Lips, and others interested in keeping wild animals safe from globe-trotting diseases.

•●•

THE AFRICAN CLAWED FROGS MAY BE OFF THE HOOK AS SUPER-spreaders because other species likely moved the fungus around too. Vredenburg thinks American bullfrogs (*Rana catesbeiana*) may have played a role in spreading disease. Because frog legs are good eating, bullfrogs, which are native to swamps and ponds east of the Rockies, have been shipped and farmed all over the world. Farming frogs by the tens of thousands is a recipe for invasion, and sure enough, in 2005 American bullfrogs joined rainbow trout in ignominy when they were listed by the World Conservation Union (IUCN) as one of the world's one hundred worst invasive species. In 2018 Vredenburg's lab found a strong correlation between bullfrog invasion west of the Rocky Mountains and the rise of Bd in the American West—precisely where the mountain yellow-legged frog and others have been hit hard by the fungus.

Like frogs, many species of salamander are exquisitely susceptible to chytrid fungi. Salamanders are also amphibians; most hatch out of eggs in ponds and streams and spend much of their adult life on land. They eat worms, maggots, and mosquito larvae, among other things. The brilliantly colorful animals are shy and slow moving, reminiscent of B-movie prehistoric reptiles: only they are not reptiles. The salamander-killing fungus is called *Batrachochytrium salamandrivorans,* or Bsal, and it too likely moved out of Asia. (Some salamanders are also susceptible to Bd.) Within a couple of years of its detection in Europe, the disease spread through Netherlands, Germany, Spain, and Belgium, killing off populations of the brilliantly colored yellow-spotted fire salamander. It hasn't yet been detected in North America, which is the world's salamander biodiversity hotspot and where a little less than half of the over seven hundred known salamander species live. Seventy-seven of those species, give or take, live in the ponds and streams of the southern Appalachian region, which makes that area particularly vulnerable to the fungus. Researchers fear incalculable loss should Bsal spread through the region. Given our lax control of disease and the demand for these exotic animals, many—including Lips and Vredenburg—worry it is a matter of when, not if, the fungi will hit the US salamanders.

• • •

THE WORLD IS LOSING ITS FROGS, AND YET MOST OF US WON'T notice. We may hear the peeper's song on a spring evening when the vernal pools and swamps warm up just enough or the wood frog's throaty cluck—though rarely do we catch either in action. Once spring passes and the mating is done, it is easy to forget that frogs are around.

But frogs in the wild are an integral part of long-formed dynamic communities. Frogs eat and are eaten, and their absence is like a tear in the food web that radiates out and impacts other species. In February 2020 a snake consuming a frog graced the cover of *Science*. The associated article was the culmination of more than a decade of work by Lips and her colleagues. The team had set up long-term monitoring sites in Parque Nacional near El Copé, Panama, to document the impact of the Bd chytrid. "We expected snakes to be affected," Lips tells me, "because so many tropical snakes feed on amphibians." So for years they turned over every plant, branch, and rock in search of frogs, salamanders, lizards, and snakes—twice a day. The results confirmed what Lips had anticipated: a loss of frogs affected snakes. Some species that couldn't manage to switch from frogs or frog eggs to some other nutrition-packed source declined. Some seemed to have disappeared completely (although, says Lips, certain snakes are notoriously difficult to find even under the best conditions). These were the "upstream" impacts on a higher predator.

Frogs are common prey, but they are also predators—and so it is reasonable to wonder what happens to the insects and snails and whatever other species they feed on when the frogs disappear. In 2020 Lips was part of another study that linked the decline of frogs in Costa Rica and Panama with a rise in malaria, which is spread by mosquitoes. "This previously unidentified impact of biodiversity loss," concluded the scientists, "illustrates the often-hidden human welfare costs of conservation failures." In places where there is little water, lakes are an oasis of life. Frogs consume what is in the lakes, and then terrestrial predators eat the frogs. In this way the food energy in the lakes flows through the frogs. When the frogs die off, there is less food for the landlubbers.

Death of any animal in large enough numbers will take its toll on the ecosystem as impacts ripple across relationships forged eons ago. The impact of fungus on wildlife species is a catastrophe most of us cannot envision and many can ignore. But when a fungus attacks a row of stately trees that stood for centuries along Main Street or turns a lush, green forest into a postapocalyptic landscape of ghostly gray snags—then we notice.

Chapter 3

CATASTROPHE

The cedar-apple rust fungal fruiting body is so brilliantly colored that if you saw it growing on a cedar branch, your first thought might be "orange sea creature" rather than fungus.

There are a lot of odd fungi, but rusts are particularly weird. They are shape shifters. Some release five different kinds of spores produced in five different kinds of fruiting bodies. Depending on their type, rust spores can travel miles when picked up by the wind or drop shorter distances if not. Some rely on insect feet and bird toes for transport. When a spore lands on a suitable host leaf, it germinates, feeds, and eventually makes its fruiting body, which will release more spores and infect more plants. Most unusually, many rusts ping-pong between two often unrelated living hosts, requiring both to complete their full life cycle. As one plant pathologist puts it, rusts are complicated beasts; they grow fine on one host but can really take off if allowed to reproduce on another host species.

Cedar-apple rust, which is native to North America, lives most often between cedar (though it also favors the eastern red cedar, which is really a juniper [*Juniperus virginiana*]) and apple trees; there are other fruiting trees it can infect too, but apple is most notable. If you have both trees nearby, depending on the season, you might have noticed the brown, pockmarked galls, hanging like ugly ornaments from the tip of a red cedar's leaves. At the very least you would have seen the gelatinous orange tendrils, one form of this rust's fruiting bodies that emerge in the spring, curling from the gall. The creature-like growths are hard to miss. The spores that are released from these reproductive structures travel when caught by a breeze. But to complete its life cycle, a spore must land on an apple or apple leaf (or related species). If it misses its target, the cycle of cedar and apple infection stops there. Apple fruits that harbor the rust may grow disfigured and stunted, and infected leaves may yellow and drop; when the infection is severe or the cycle continues, the tree may stop producing. But both the apple and the red cedar tree usually survive. Other rust fungi are not so forgiving.

White pine blister rust, *Cronartium ribicola*, grows on some of the white, or "soft," wood pines; like other rusts, the fungus requires a different plant, in this case currant and gooseberry bushes—shrubs in the genus *Ribes*—to complete its life cycle. Though there are *Ribes* species native to North America, the most problematic plant, *Ribes nigrum*, or black currant, hails from Europe and Asia, having crossed the Atlantic with colonists who settled around Massachusetts Bay. The fungus will do little harm to the currant bushes, but it can kill the pines. Many of these softwood pines are valuable forest and timber trees, and for over one hundred years foresters have tried to control damage from the fungus with varying degrees of success. Now forest researchers and managers fear that white pine blister rust is driving populations of one of North America's most iconic and wildest pines, the whitebark pine, toward extinction.

Whitebark pine trees (*Pinus albicaulis*) are both a keystone and a foundation species, providing opportunity for a diversity of plants and animals and the promise of ecosystem stability. They live on steep, rocky slopes at high elevations where many other tree species cannot grow, setting their roots into soil and cracks along steep slopes and surviving altitude, wind, and ice. At treeline they may grow as a solitary tree clinging to an exposed rock face, or they may form tree islands—isolated clumps of low-growing conifers providing food and shelter for a community of plants and animals in an otherwise harsh landscape. Depending on where they are growing, they are also host to a varied understory of plants, including sagebrush, whortleberry and huckleberry, bear grass, sedge, and heath. Whitebark pines live in western subalpine, temperate forests, knocking branches in some regions with its relatives like the limber pine, or spruce, fir, hemlock, alpine larch, and Douglas fir. Whitebark pine cones produce remarkably large seeds compared with other pines, and the seeds are rich in fat. Red squirrels cut cones and cache them for winter. Some of those will be stolen by both grizzly and black bears, which also rely on the calorie-rich seeds. The most important consumer of whitebark pine seeds is a bird called the Clark's nutcracker. Like squirrels, the birds cache their seeds. But they bury more than they can eat, and given

WHITEBARK PINE AND CLARK'S NUTCRACKER

the right conditions, some of those seeds will germinate. This is how whitebarks are propagated. The birds are not solely dependent on the trees to survive, but the trees rely on the birds to reproduce.

If you look at a map of whitebarks in the American West, you will see two streaks ranging from the coastal ranges of British Columbia south through the Cascades and down through the Sierra Nevada; to the east is another smear of whitebarks running from Alberta, Canada, and British Columbia down through the Rocky Mountain ranges in Montana and Wyoming. These trees populate our high-elevation national parks, including Yosemite, Crater Lake, Glacier, and Yellowstone, and are important members of the three largest wilderness systems in the West. If you have hiked these regions, you have likely encountered, maybe even admired, these often solitary trees.

Whitebarks embody forest untouched by humans, nature's spirit. They are the gatekeepers separating the civilized from the wild, tough old trees that have survived centuries of weather, insects, and disease. Scientists now worry that white pine blister rust is one challenge too many.

●●●

THE WHITEBARK'S STORY OF DECLINE BEGINS WITH ITS COUSIN, A majestic tree called *Pinus strobus*, the eastern white pine. Both are members of the five-needle pine clan, a grouping of trees whose needles grow in bunches of five. All pines are conifers, members of an ancient group of trees called gymnosperms, which makes them distant cousins of prehistoric holdovers like ginkgo trees and fern-like cycads. Redwood trees are conifers as are bristlecone pines and hemlocks. Most conifers acquire their energy through evergreen needle-like leaves, while some, such as the junipers, have leaves coated in tiny reptilian-like scales. They make no flowers, unlike maples, oaks, and other flower-producing trees, relying on seed-bearing cones instead. Male cones make pollen while the females, once fertilized, make seeds. Of the over one hundred species of trees recognized as *Pinus*, nine five-needle pine species are native to North America. The only one of those in the East is the eastern white. The western white, fox-tail, sugar, limber, two bristlecones, southwestern white, and white-bark are all western trees. Many of them grow under conditions other trees could never tolerate. Eastern white pines favor the cold but not high altitudes. Typically, these pines grow in forests of hemlock and northern hardwoods from the Northeast coast, through the Great Lakes region, and down the Appalachians toward northern Georgia. They are sometimes the first trees to begin creeping into abandoned fields where there is an opportunity to transition back to forest.

If you find a rare white pine elder in an eastern forest today, you will notice it is often remarkably tall and often straight with thick, brick-like bark. The largest remaining trees are nearly 20 feet (6 meters) in girth and soar to over 160 feet (50 meters), the height of a 10-floor building. Some are hundreds of years old. Most but not all grow quickly as a single, unbranched trunk reaching skyward. The Iroquois refer to the white pine as the Tree of Peace: the five-needle bunches symbolized cohesion among what were at one time warring nations; its evergreen needles represented a lasting peace. To imagine a forest kingdom of these quiet giants is to imagine a time before the Europeans and before the logging frenzy.

Historically, loggers prized eastern white pines for ship masts and lumber, and cut them one by one, ten by ten, then by the hundreds and millions. The colonial British navy ran their sails up white pine masts cut from New World forests; broad pine floorboards ran across homes and mansions; and figureheads—eagles or busts of women carved from the old growth so-called "pumpkin pine"—decorated sailing ships' bows. In 1691, to ensure that colonists did not take the best trees for themselves, the English crown issued a proclamation: the largest of these trees that were not on property granted to any private person, those "of the growth of 24 inches in diameter," were to be marked property of the king. The restriction contributed to a growing discontent between those who wished to supply the mother country and those who wished to build a new one. As the supply of old growth and native trees dwindled, the British attempted to grow their own white pines by sending seeds back to England. They grew but never achieved the stature of the North American stock.

Centuries later, after forest turned into field and field returned to forest and nineteenth-century loggers became increasingly efficient at cutting down trees, regional forests became depleted of the highly valued white pines. Replanting efforts followed, but the sheer number of seedlings needed to cover thousands of pine-depleted acres across New England was beyond the capacity of local nurseries. Ironically, importing seedlings of this New World native tree from nurseries in Germany, France, and Holland was relatively economical. Europeans had plenty of experience as nurserymen, managing young trees grown from seed. And so, for a time, foresters and nursery owners in the States began importing white pine seedlings by the millions from Europe. In 1909 alone, a single nursery in Germany sent several million pines for planting to hundreds of locations in the United States. But there was more to the trade than the young white pine seedlings. Some of those young trees arrived infected with white pine blister rust.

• ● •

IN 1905 A FUNGUS SAMPLE COLLECTED FROM A PLANT NURSERY NEAR Philadelphia landed on the desk of US Department of Agriculture (USDA) mycologist Flora Patterson. The sample had been collected from a young white pine and was identified by Patterson as "a *Peridermium* which causes what is called white pine blister rust." A year later the same fungus popped up on currant bushes in Geneva, New York, and four years later, in 1909, a young forester also in Geneva noticed some odd swellings on white pine stock that had shipped from Germany.

The fungus was already well known in Europe, where it infected native pines but didn't kill them. Familiar with the fungus and knowing the devastation it could cause in the North American trees, a German forester named Carl Schenck had warned that importing European pine stock could lead to disaster. Because the fungus can incubate unseen for two or three years before making more visible spores, it made identifying the fungus particularly challenging. When more samples of the fungus collected from snippets of currants, gooseberries, and pines began trickling into Flora Patterson's office, she began to worry.

Like most fungi, blister rust fruiting bodies send out plenty of spores. The full life cycle typical of rusts is bizarre. Rusts are related to mushrooms, and like mushrooms at one point in their life cycle, two different genders of spores will join together and have fungal sex. When a rust spore lands on the needle of a five-needle pine and if the weather is to its liking—cool and misty—the fungus by way of basidiospores begins its invasion, entering the thin, waxy needle through the stomata. Plant leaves are dotted with these pores that open and close, taking in gases like carbon dioxide that the leaf turns into food while releasing oxygen as a waste product. Stomata are like an open door for rust, and once the fungus is inside, slender hyphal threads begin to probe the leaf for food. Each basidiospore will grow into a colony of a particular mating type. Some refer to them as "+" and "−": two opposite types. As the fungus grows, mycelia spread down the needle and into the stem. The needles turn rust red. A pine branch waving clumps of dead, red needles is a sign of infection. Eventually the fungus will reach into the branches and trunk. Seedling, sapling, or maturing, grand old tree—rust will invade. A few years after the initial infection, in the spring as the temperature warms, blisters erupt through a canker in the infected tree's bark on its branches or trunk and produce spermatia (or pycniospores). Some will be released in a sticky liquid and may be carried off by insects or other species or by rain. Some will germinate on the tree into hyphae; when hyphae of opposite mating types (the + or −) meet, they fuse, allowing their genetic material to meet and recombine. A year or so later the fungus again emerges from the trunk and begins releasing powdery, yellow-orange aeciospores. These spores have evolved to travel with the wind. They will land on soil or touch down on the shoulder of a hiker or the windshield of a car. They will land on rocks and streams—and on hemlocks and fir trees. Some will land on the leaf of a currant or gooseberry bush. Rust enters the leaves through the stomata, just as it does in the pine. Once infected, a leaf might turn yellow and fall to the ground, but the plant will soldier on.

A couple weeks later the underside of infected leaves will appear as if dusted with a rusty powder. These so-called urediniospores will spread infection to other *Ribes* leaves. The plants become like rust bioreactors, allowing the fungus to infect more leaves and more plants and to multiply. When temperatures cool and day length shortens, hair-like telia—a different spore-making structure—grow from the leaf's underside. Telia release basidiospores that can only germinate on the needle-like leaf of a pine. These spores are more delicate than aeciospores and are carried by the wind across a shorter distance, perhaps many hundreds of feet. Insects, birds, and little mammals may also carry rust, moving spores trapped between their toes or stuck to fur or feather. Once infected, trees, like any other living creature, are not defenseless. Sometimes these defenses are sufficient, sometimes not. White pine forests, both wild and managed, were full of naïve hosts that, it turns out, were largely unable to defend against the new exotic fungus.

Like humans, trees are built of cells, with different kinds serving different roles. As a tree grows, some cell types radiate horizontally while others stack one atop the other. Some become conduits where nutrients and water flow up and down. In the tallest trees molecules of nutrients and water travel hundreds of feet from roots to shoots to leaves, and sugars from needles travel down. Some tree cells remain alive for decades. Others grow into roots holding the ground and communicating with kin and neighbors through a network that includes an entire underground realm of fungal mycelia and other microbes. There are cells that grow into leaves that make food and are eventually shed, or into fruits, nuts, and cones. When certain cells die, they remain a part of the tree, providing structure, strength, and protection.

Trees also have multilayered defenses from illness and invaders. Bark, formed from layer upon layer upon layer of dead cells, is a first defense. It is sufficiently protective that some fungi require nicks and cuts or the rooting around of birds and insects to gain entry. Once that barrier is breached, trees, just like other plants, have chemical defenses, some of which have evolved to deter microbes, insects, and other predators. But when these defenses fail, or if a tree loses a branch or becomes wounded by animals or disease, it is unable to repair this damage as some animals might. There are trees within trees within trees, as if each concentric ring is its own Russian doll–like tree. Even if it must sacrifice a limb or portion of the trunk, a tree grows and makes new cells.

When fungi infect the inner bark, the tree responds with a canker: a bulging growth circumscribing a wound that is sunken into the trunk or branch. The cankers that form on infected trees help compartmentalize the infection. Sometimes, if a pathogen continues to grow, the canker forms concentric rings: canker over canker over canker. When you see these bulging imperfections in a tree's trunk or on a branch, you are seeing a struggle for survival. Often the tree lives on, albeit more gnarly and scarred than it once was.

Older trees, with more options for walling off, or shedding, infected branches, may have a better chance of surviving fungal pathogens and other insults than younger trees. As the fungus and the cankers grow, eventually the flow of water and nutrients may become blocked, effectively strangling the tree to death. This is how blister rust kills: first the needles, then the stems, the branches, and finally the trunk. Young pines that have fewer parts to sacrifice are more likely to die and will succumb more quickly. Despite the trees' best defenses, in terms of the tree's evolutionary history, blister rust was a novel challenge, and many individuals were not adapted to survive this new enemy.

From the mid-Atlantic to Canada to Maine, orange pustules dotted the branches and trunks of eastern white pines, and red needles waved from infected branches. In the West the rust again arrived from Europe, first seen in British Columbia sometime in 1910 and then spread to forests in Washington and Oregon. By the early 1930s it reached Northern California and soon after the Rocky Mountains. The deadly fungus swept through western forests thick with towering sugar pine and western white pine, and whitebarks, too.

••*

WHEN RUST BEGAN ITS RUN THROUGH THE NATION'S WHITE PINE forests, managers at first tried pruning away infected branches, but the fungus continued to spread. Alarmed at the high mortality rates and worried that the forests and their commercial uses were doomed, foresters and scientists turned to a different disease control strategy: breaking the rust by removing its alternate hosts, the black currants and other *Ribes*.

The shrubs grew everywhere. In addition to the imported black currants, native *Ribes* were plentiful too, including the odd grape-like gooseberry, although the flavor-packed imported berries were most popular for cultivation. By 1916, a few years into the epidemic, laborers were pulling up *Ribes* plants in Massachusetts as an experiment in control. Studies suggested that eliminating plants two to three hundred yards from the nearest pine could reduce disease. The estimated cost of removal was forty-two cents per acre, about the price of a dozen eggs. In Maine workers were paid thirty cents an hour; elsewhere others were paid in groceries or clothing. Some

regions paid by the acreage cleared. In the 1930s, in the middle of
the Great Depression, the Civilian Conservation Corps hired able-
bodied men for the job. In the late 1930s and early 1940s as men
went to war, high school boys filled in. Locals who favored the pines
were all for destruction of the berry bushes, but some, particularly
those who harvested the fruit, harassed workers and asked to be
paid for their loss or complained about government overreach. But
by 1965, in the Northeast, more than twelve million acres (with
white pines growing on five million of them) had been cleared of
Ribes. Blister rust remains but is no longer the threat to the eastern
trees it once was.

The West was different. *Ribes* grew on mountainsides, in deep val-
leys, and in other remote locations far from settled towns and villages.
Workers there would hike into the mountains and settle into camps
spending months at a time clearing the shrubs. "I was first introduced
to Blister Rust Control in 1953," writes Gerald Barnes, a retired forester who spent much of his career working with western pines and
blister rust. He had just graduated from high school in Grants Pass,
Oregon, when he joined an eradication crew working the Siskiyou
Mountains region. It was rugged wilderness work. Laborers camped
and then walked for miles, at the mercy of rattlesnakes, bears, steep
cliffs, and extreme heat. They cleared *Ribes* using hand-held hoes,
turned the bushes upside down, and clipped or poisoned the plants'
roots. Along some large creeks *Ribes* were sprayed with herbicide.
Workers walked the creeks carrying military-issue packs full of herbicide (2,4-D) and mixed the spray as they needed. "It was a bushy
and messy job," writes Barnes, who stuck with the program until the
summer of 1958. By the time the national program ended in the late
1950s, millions of acres had been cleared by some ten thousand workers. Despite the success in the East, rust remained a problem in the
West because it was so vast and the terrain so rugged.

New Hampshire was the first to ban planting of *Ribes* bushes in 1917. A federal ban followed, prohibiting import and cultivation of *Ribes*. As the fungus was somewhat controlled in the East and breeders began producing rust-resistant varieties, growers pushed to end the bans. The federal ban on *Ribes* imports ended in the 1960s. But because states can make their own rules, some continued to ban black currants. The result is a confusing patchwork of what plants can grow where. In Massachusetts, as in Maine, growing black currants is still banned. The state sits between both Vermont and Connecticut, neither of which bans the shrub. For a while New Hampshire encouraged growers to try a rust-resisting currant strain called Titania. But when white pines growing near some of those bushes became infected and a study found that rust had overcome the strain's resistance, the state banned the plants. There are no restrictions on buying and planting *Ribes* in Oregon, Wyoming, Washington, and other western states. And yet blister rust remains a major problem in all these regions.

In some parts of the northwestern United States and southwestern Canada, where rust has settled in, almost every standing whitebark is infected. Where there were once beautiful, high-elevation forests, the rust has left gnarled gray ghosts. Well over half the cone-producing whitebark pines have died across the western United States, with even higher losses in the northern Rocky Mountains. When such large swaths of trees are killed, cone production is reduced, and scientists fear that wildlife that has coevolved with the trees, including the Clark's nutcracker, will abandon them for other food sources. If they do, it could be disastrous for the surviving trees.

For more than forty years Diana Tomback, an ecologist at the University of Colorado, Denver, has studied the relationship between the Clark's nutcracker, a handsome gray bird with striking black and white wings and tail, and whitebark pines. "They are *so* dependent," she says of tree and bird. The whitebark's deep purple cones grow on

WHITEBARK PINE GHOST TREE

the tips of the highest branches on the crown of the tree. The seeds, which are not sown without some assistance, do not have "wings" like some other conifer seeds that can be propelled by the wind and disperse. Instead, the large seeds stay locked in cones that remain closed, even when ripe. In the fall, when the birds need them most to prepare for the winter, the cones, which are easy to access on the treetops, are packed with seeds. Nutcrackers grasp a seed with their long, strong, and slender beaks, and with a slight upward jerk the seed enters an opening under the birds' tongue and is stowed in the sublingual pouch, which when full, puffs out like a bullfrog. The birds cache a few or dozens of seeds at a time in soil or just under the forest litter; in a good cone year each bird will hide tens of thousands of pine seeds. In the spring nutcrackers feed their young with seeds that haven't been stolen by rodents or other marauders. There are a lot of ways to lose the seeds, so storing excess in hundreds if not thousands of caches is insurance. It also means that sometimes uneaten seeds can germinate. And that is how the whitebark pines move across the landscape or reforest after a wildfire or populate the treeline. Buried for a couple of years or more and given the right conditions, a cache of seeds will sprout, sending up a cluster of seedlings. If the nutcracker were to abandon the trees, there would be little hope for new whitebarks. The birds help propagate the trees, and the trees ensure there will be food for the birds. Both benefit from this tight relationship. But blister rust has disrupted the cycle, whitebarks are dying, and the birds can find seeds elsewhere. Tomback worries that if something isn't done, some regional populations of whitebarks may be lost within a century.

The fungus is not the tree's only threat. The changing climate is reducing suitable habitat for the trees, forcing them to move up in elevation into cooler habitats. And some regions where the trees now grow will probably heat up in the near future. Add to this the natural boom and bust cycles of beetles. Mountain pine beetles are native to the West; the lodgepole pine and other pines are their favorite host, but they infest all pines, including the whitebarks. A female will chew her way through the protective outer bark and into the softer, living inner bark. As she does, she releases pheromones, chemicals that attract other beetles—both male and female—to the same tree. There she lays dozens of eggs. The immature beetles spend most their lives beneath the bark, feeding from the tree. When they are mature and ready to mate, adult beetles emerge from one tree and within days chew their way into another. A severe attack will kill the tree. A massive outbreak in a region kills millions of trees. The warming trends from climate change seem to favor beetle infestation. As blister rust mingles with the direct and indirect impacts of climate change and the beetles, the whitebarks face an extraordinary triple threat, which is most disturbing because fungus alone was enough to drive a different tree species, the American chestnut, into virtual extinction.

.●.

THE AMERICAN CHESTNUT (*Castanea dentata*) ONCE POPULATED THE
spine of the Appalachian Mountains, from the Blue Ridge Mountains
to the Berkshires and beyond. Most of us have never seen a full-grown
American chestnut, but there was a time when it ruled. These towering trees with cottony white catkins dominated the forest canopy in
summer and dropped bucketloads of sweet nuts in the fall. There are
reminders all around: Chestnut Plain Road, Chestnut Hill, Chestnut
Street. The fast-growing hardwoods meant jobs for lumberjacks and
rot-resistant lumber for homes, ship masts, and railroad ties. In the
early 1900s the chestnut timber industry was worth more than $20
million (which would be over $600 million today). Some trees were
enormous, measuring well over ten feet in diameter; at least one large
chestnut provided enough wood to build a whole cabin. There were
chestnuts for roasting and odes to nuts and photos of families gathered around outsized chestnut trunks. Like whitebarks, chestnut trees
also provided for wildlife. The nuts were so plentiful that farmers
sent their pigs into chestnut forests to fatten up and later feasted on
chestnut-flavored pork. The trees were also foundational, even if in
a more human-oriented system. In some regions, scientists estimate,
every fourth tree was a chestnut. By another estimate nearly 30 percent of Great Smoky Mountain National Park was once thick with
chestnut stands.

Among the thousands of trees and shrubs growing in the New
York Zoological Park, now known as the Bronx Zoo, were chestnuts.
The park, a project of the New York Zoological Society, was a planned
environment, designed to create a nature-like sanctuary in the middle of a growing city. The zoo opened its gates in 1899. The animals
were exhibited in more naturalized habitats rather than in cages, and
enclosures—when they existed—were barely visible. There was a
herd of antelope rather than an individual animal, along with prairie

dogs and beavers. In the middle of New York City there was Mountain Sheep Hill and an otter pool, crocodiles and alligators, all surrounded by forest. A key to the park's success was the greenery, and a young German immigrant named Hermann Merkel, the zoo's first forester, was responsible for all of it. If it grew and was green, it was Merkel's. As the city expanded, the forests within became increasingly precious. A good swath of them were growing in the zoological park. The trees, reckoned the Zoological Society, were "not even second" to the animals that would soon be living there. Animals could be replaced, but a centuries-old oak or chestnut, should it fall, would be gone forever. "The death of a large forest anywhere within the limits of New York City," noted the society in 1898, "is nothing less than a

AMERICAN CHESTNUT TREE

calamity." And so they hired Merkel to preserve the oaks, elms, tulip trees, and chestnuts that shaded and sheltered the park grounds.

Hermann Merkel knew that park. In addition to maintaining the health of thousands of trees, he and his crew planted grasses, shrubs, and flowers. When he wasn't tending to the plants, he could be off helping to capture escaped animals (a small puma, "no more, no less" dangerous than a cross fox terrier, and burrowing prairie dogs on the loose). He built and fixed rock walls by the Lion House, ensured proper drainage by the Antelope House, and constructed concrete barriers around Prairie Dog Village. And so in the summer of 1904 he couldn't help but notice oddly yellowing leaves and how they curled up around the edges, as if it were fall, on some of the park's chestnut trees. When he looked more closely, Merkel noticed that the limbs holding those dying leaves were dotted with red and orange spots, pustules suggesting some kind of fungus—obviously not good, but only a handful of trees seemed to be affected. He wondered if it was possible the trees had been particularly stressed that year since there had been a cold winter followed by drought. Maybe whatever the parasite was, it would die over winter and never return. But it did return.

By the next summer, in 1905, almost every one of the park's 1,400 chestnut trees was dead or dying—virtually every single chestnut tree in the park. It didn't matter where they were, how old, or how large. From recently planted sprouts to a few "primeval forest"–sized trees with trunks ten to twelve feet around, all were infected. It was the calamity of forest death that the park leaders had feared. Hoping for a cure and some help, Merkel sent a sample to the USDA, where it landed on Flora Patterson's desk. She unfortunately misidentified it as a common fungus of certain trees (though she noted that she had not known it to infect chestnuts). Pruning back the diseased branches and treating with a new fungicide was the recommended fix.

In the early 1880s the French botany professor Pierre-Marie-Alexis Millardet noticed something odd in a vineyard in Médoc (a region of Bordeaux), parts of which were infected with a powdery mildew. At the time, some growers treated their vines with a mixture of copper sulfate and calcium hydroxide intended to scare off anyone wanting to poach their grapes with its unnaturally green-tinged residue. But Millardet saw that the treated grapes appeared free from mildew and wondered if the deterrent worked as a fungicide as well. It did—and without compromising the grapes or the plants. At least that's how the story went. It is likely Millardet already knew the chemical mixture could prevent fungal infection. A year later word of the "Bordeaux" mix crossed the Atlantic and was popularized by Patterson's boss, USDA plant pathologist Beverly Galloway. It became the go-to fungicide for crops in the United States, and it remains a popular "organic" fungicide today. But there's a catch. Bordeaux mix can't penetrate plants. Fungi are killed when the residual copper on the leaves is wetted with dew or rain. The moisture allows for the release of copper ions, which can disrupt proteins including essential enzymes in fungi and other pathogens. It is a topical rather than a systemic treatment and works best when fungus is exposed on a plant's leaves or stems or bark. Once applied, it easily washes off in the rain and is no longer effective on the plant (and there are problems of copper accumulation in soils where plants have been treated for decades). On chestnuts the mix would have been useful against blight *if* the fungus was caught before any spores germinated. Once the fungus invaded the tree, the topical treatment was useless. Even if it could have worked, the problem of spraying the limbs and trunks of the towering trees with enough fungicide to control the disease was practically insurmountable. Merkel began pruning and poisoning, but with over a thousand trees throughout the park and word that even more trees beyond the park were infected, the impact of his work was underwhelming. This was no ordinary tree fungus. Merkel decided to seek a second opinion.

William Murrill was the new assistant curator over at the New

York Botanical Garden, not far from the zoological park. Merkel asked if he would look at the ailing trees. Young and ambitious, Murrill disagreed with Patterson's diagnosis. Even if the USDA had correctly identified it, why would it suddenly turn so lethal? The mycological mystery was a career opportunity for Murrill, as he wrote in his autobiography (referring to himself in the third person). It was "just another timely round in the ladder of luck he was climbing to fame and influence." He brought some fungus back to the lab and infected chestnut twigs to first confirm that the fungus caused the symptoms.

Decades earlier, microbiologist Robert Koch, seeking to set standards for identification of invisible yet deadly diseases, developed a set of steps for linking microbes with disease. His discovery rested in large part on the seminal work of the chemist and microbiologist Louis Pasteur. In the 1860s Pasteur conducted his now iconic experiments on the role of microbes as cause rather than consequence of disease. Pasteur demonstrated that microbes were not only capable of causing disease, but in some instances they were required. Following Pasteur, Koch began isolating disease-causing organisms. But how to ensure they were necessary for disease? Koch's contribution to what is known as the germ theory of disease was a reproducible strategy for linking cause and effect: isolate the suspected cause (a particular microbe) from its disease host (one should not be able to isolate the microbe from a healthy host), infect a healthy host, observe disease, and reisolate whatever was causing it once again. Like Koch, Murrill isolated, infected, and isolated again. Young chestnut twigs bloomed with fungal pustules just as they did in the local trees. Within several weeks spores oozed from the chestnut saplings he had infected. Murrill had successfully linked a novel deadly fungus with a disease. He named it *Diaporthe parasitica* and sent an infected twig to the USDA to claim his discovery. The fungus would later be renamed *Cryphonectria parasitica*, or chestnut blight.

When a fungal spore lands on chestnut bark and germinates,

mycelia enter through a wound or a nick and then grow beneath the outer bark. Unlike white pine blister rust, blight can be quick to reproduce, sending out spores within weeks of infection. Usually the asexual spores are sticky and ooze from the fungal fruiting bodies, whereas sexual spores go airborne, either shooting out into the air or spread by the wind. When spring arrived at the zoo, millions of asexual spores called conidia would have squeezed out through brilliant sulfur-yellow tendrils called pycnidia. The sticky spores clung to the feet of insects and birds like nuthatches, brown creepers, and woodpeckers. Droplets of rain might have splashed some spores onto lower branches or nearby trees. When a male spore from one fungal strain and a female spore combine, a *new* kind of sexual spore is formed called an ascospore. Packaged into a bulb-like structure, ascospores can shoot up, out, and into the air. When caught by the wind, ascospores from the zoo and elsewhere landed on chestnuts farther afield, broadening the infection radius. Given the expediency of the fungus and its rapid invasion of its host, it is not surprising that neither fungicide nor branch cutting could control the beast.

"It is safe to predict," wrote Merkel in the zoological society's 1905 annual report, "that not a live specimen of the American Chestnut will be found two years hence in the neighborhood of the Zoological Park." He was right. By 1910 the botanical garden too lost over one thousand chestnuts. The fungus moved along the hillsides and down into the chestnut forests of the Appalachians. The fungal front was moving at a clip of some twenty-five to thirty-five miles a year, killing chestnuts from Pennsylvania to Georgia. Those that were not killed by blight were felled by foresters urged to cut the trees while they still could.

This widespread destruction of a key forest tree was frightening. Drought, fire, insects might destroy a forest, but at the time, these were natural disasters; they had happened before. Native insect populations like the mountain pine beetle rose every so often on their own cycles. Many trees and wild plants need fire to regenerate. And even crop-killing fungi had been around since biblical times. But a fungus that wiped out a single species of tree was a new thing. Magnificent chestnuts along hills and mountainsides turned into ghostly gray dead wood. The *New York Times* described the blight as "wickedness in the tree world." Letters by the hundreds from as far as Virginia indicated how quickly blight was moving down the East Coast. Some suggested that the blight was "a scourge for sinfulness and extravagant living" and was a call for prayer or maybe even a revival. In places where American chestnuts were particularly thick, their sheer density enabled the spread of fungus from tree to tree.

Ever hopeful that humans could beat the fungus, in 1912 James Wilson, the secretary of agriculture, wrote, "There is no contagious disease known that does not yield to sanitation and quarantine." But then neither Secretary Wilson nor anyone working at that time had witnessed a killer fungus on the loose in the forest. The most uncontrolled and uncontrollable kind of outbreak, the disease was unprecedented. By some estimates three to four billion chestnuts died within a few decades, forever changing the forests, the culture, and livelihoods. When Hermann Merkel died in 1938, all but a few chestnuts (those trees that had been planted far across the country or that happened to grow in isolated pockets of forest and so were essentially quarantined) were dead or dying.

Several years after the discovery of the fungus in the Bronx, scientists reported that it had been found living on a chestnut tree in China, where it most likely originated. Like white pine blister rust, this plague also arrived by ship in the stem of an import.

• ● •

AT THE TURN OF THE NINETEENTH CENTURY A HOMEOWNER OR orchardist could buy a Spanish or American or Japanese chestnut by mail order from nurseries in New Haven, Connecticut, or Rochester,

New York, or Biltmore, North Carolina, for around a dollar or less. Americans could also find fruits and nuts grown from imported plant stock in local markets. Farmers could grow wheat from Russia or kale from Croatia. As with animals entering the country today, novelty was of more interest than the diseases a new plant might carry. The job of gathering seeds of fruits and grains, vegetables, novel trees, and other ornamental plants then as now fell under the auspices of the USDA. Plants from around the world could turn the nation's soil into money—good for consumers, farmers, and the economy. The problem of imported pests and pathogens was well known at the time, and the USDA employed mycologists like Flora Patterson and entomologists and other scientists who could identify and then provide advice and solutions to farmers and landowners. But it had little interest in *preventing* problems. New technologies, better breeding, and chemistry could almost always triumph over nature, or so the thinking went. Over dozens of years, under the auspices of the USDA, tens of thousands of plants, seeds, and saplings collected from far-flung lands were crated up and shipped across the sea to set root in American soil.

One of the most prolific collectors was David Fairchild, who in 1898 became the first director of the USDA's Office of Seed and Plant Introduction. He traveled through Europe, crossed the Suez Canal, and steamed off to Java and Sumatra. Along the way he collected edible plants and seeds. While at the USDA, Fairchild oversaw the introduction of pistachio trees, cashews, flowering cherry trees, lemons, nectarines from what is now Pakistan, and more than a hundred thousand other food plants. Under his watch, seeds, plants, and clippings of crop plants and ornamentals for the garden were shipped by agricultural explorers who traveled to remote villages from Russia, China, Japan, Algeria, and dozens of other countries. Fairchild and his colleagues changed both the way the country ate and how Americans planted around their homes. By the end of the century hundreds of thousands of plants had taken root far from their native lands. Americans enjoyed the variety, and farmers benefited from the crop diversity. The great influx overshadowed the dark side of the USDA's effort, the insects, rusts, scabs, and other diseases that hitched a ride to the continent. There were a few critics of Fairchild and his mission, and one of the most vocal was Charles Marlatt.

Marlatt had known Fairchild as a kid; both had grown up in Kansas, and Marlatt, six years older, considered Fairchild a sort of younger brother. Fairchild gravitated to plants, while Marlatt studied insects and crop pests. They were kindred spirits, representing the yin and yang of modern agriculture. Fairchild shipped out to exotic lands, beginning his itinerant career as an agricultural explorer, while Marlatt studied periodic cicadas and crop pests in California, Virginia, and Texas. In 1889 Marlatt was hired as an assistant at the USDA's Bureau of Entomology. By the time the two scientists reunited in Washington, DC, they had remained good enough friends for Marlatt to stand as best man for Fairchild when he married Marion Bell, Alexander Graham's daughter, in 1905. The friendship would be short-lived.

Marlatt considered the edible and vegetal wonders imported by Fairchild and others as disease-infested Trojan horses, each a potential catastrophe for American farmers and forests. During his tenure at the USDA he sought to prevent the import of pests and pathogens; it was an unpopular stance, which his predecessors at the agency had also championed but to little real effect. The department was bent on expanding the nation's agricultural prowess, and the nation's citizens enjoyed the novelty of exotic fruits and vegetables and plants and trees for their gardens. For over a decade Marlatt had worried about all the unseen things that could destroy a crop or turn a healthy forest into a ghost forest, while Fairchild could only find the good in a better strain of peach, apple, or grain variety. The two old friends, each wedded to their belief about the good and evil of imported plants, had turned against one another.

In 1909 Marlatt began pushing for federal protections against imported pests and pathogens. Ideally, he would have liked to end the collection and importation of exotic plants altogether, but instead he drafted a bill that would give the USDA the authority to control how imported plants were received and distributed. But the Committee of Eastern Nurserymen, represented by a small group dependent on imports, pushed back. The nurserymen, Marlatt later wrote, "careless of the consequences of the country at large, feared some slight check on freedom of their operations." They were joined by the ladies' garden clubs, whose members numbered in the thousands. These were educated women who had been excluded from horticultural societies and botany clubs. They raised funds to beautify landscapes and were active in the politics and preservation of the American landscape. That the position of these women against the legislation held even more influence over Congress than the nurserymen's, wrote Marlatt, "was very ominous for the bill." The effort failed. Then the cherry trees arrived.

On January 6, 1910, two thousand Japanese cherry trees were sent from Japan to Washington, DC, for planting on the newly created mall around the Washington Monument. The trees had been Fairchild's idea. Years earlier he and his wife, Marion, had planted Japanese cherry trees on their own property and noticed the joy the fluffy pink blossoms brought to visitors. Wanting to share the beauty, Fairchild imagined a botanical scene free to anyone and everyone: flowering cherry trees planted along the new Washington Mall. The timing was fortuitous. In 1907 the United States and Japan had entered into what was known as the Gentleman's Agreement. The Japanese agreed to limit immigration to the United States. In turn, the United States would request that the city of San Francisco, where racial intolerance was on the rise, rescind its recent order to segregrate "all Asian" children into separate schools. For President William Howard Taft the trees would be akin to a handshake following the deal, a way to tame tensions. Fairchild arranged for the importation of three hundred trees; the mayor of Tokyo selected and sent two thousand. It seemed as though everyone was happy—except for Marlatt.

Because the trees arrived under the auspices of the USDA, Marlatt saw an opportunity: he had the authority to inspect the imports. The cherry trees arrived in Seattle in December 1909 and were sent on by rail to Washington, DC. They arrived at what was to be their final destination in January 1910. Marlatt sent a crew of entomologists, including mycologist Flora Patterson, to inspect the trees. They reported finding scale insects, wood-boring larvae, crown gall (caused by a bacterium), and a fungus that could only be identified to its genus. After the inspection Fairchild wrote, "I found myself in a hornet's nest of protesting pathologists and entomologists." Marlatt recommended that the trees be destroyed. A country that legally protected agriculture, he wrote, would not have allowed the importation of such diseased trees. The report was sent to Taft, who ordered that the trees be burned. On January 28, the cherries, which had traveled across the Pacific and then across the continent, were piled together and torched—not a good look for relationship building. A *New York Times* editorial worried about "wounding to Japanese Sensibilities," when the "pretty gift of root and branch" was burned. Perhaps, wrote the *Times*, a "carefully arranged accident" would have been wiser. But there was no international incident, and as the nation worried, the mayor of Tokyo offered profuse apologies for sending defective trees. Two years later more than three thousand young, healthy cherry trees arrived from Tokyo, passed inspection, and were planted along the Tidal Basin. The Japanese ambassador's wife had the honor of planting the second tree in DC soil.

In 1911 Marlatt, still seeking legislation after a very public display of the risk from imported trees, took his argument to the people, publishing an article in the widely read *National Geographic* (a society magazine whose president at the time was Alexander Graham Bell,

Fairchild's father-in-law, and that included Fairchild as an associate editor). Marlatt used this opportunity to shake up readers, frightening them with a litany of imported pests and pathogens running amok in their homes and yards and across the nation. He included images: masses of gypsy moths on a tree trunk, black tumor-like masses of potato wart clinging to the roots of a potato plant (the disease, already in Newfoundland, threatened the US potato industry), swaths of dead chestnuts killed by the blight. "The entire chestnut timber of America seems to be doomed," he wrote, adding that "all this might have been saved with proper quarantine laws." He noted that 50 percent of the known pests were of foreign origin. Other countries had already recognized that imported plants came bearing gifts of disaster. Some had strict quarantine and inspection laws; others outright banned nursery imports from the United States. The nation was, he wrote, the only "great power" without any import regulation. As such, the country had become a "dumping ground" for all the infested plant stock rejected from European ports. France had already experienced a devastating infection caused by a tiny insect called *Phylloxera vastatrix*. The pest, which fed on grape leaves and roots, was native to the Mississippi River Valley. In 1862 it was exported to France along with some grapevines and soon took hold in Provence and Bordeaux and was spreading toward Burgundy. In time the insect-infected vines appeared in Spain, Italy, Germany, and beyond. Pesticides of the day couldn't kill the bug. Desperate growers, realizing that European vines could no longer sustain the onslaught, tried growing American grapes or hybrids. But to the French the resulting wines lacked good flavor. Reluctantly growers took to grafting their vines onto American root stock. While recovering from the infestation, France began importing wine; impostors began selling cheap fakes; and the pest provided uninfected countries with an opportunity to increase their exports, at least for a time until French vineyards rebounded. In response to the devastation, in 1878 several European countries banded together and created a phytosanitary agreement requiring certification of *Phylloxera*-free imports. The United States, for the most part, remained open for business with little concern. Several months after Marlatt's article, Fairchild responded, also publishing in the *National Geographic*. He included images of fig trees and mangos and a photo of sprawling alfalfa fields and rows of seedless grapes that could be used to make seedless raisins—found in Italy. There was also a photo showing burlap bags full of seeds and stock. In short, instead of death and destruction, there was abundance to be had—if only it could be collected and brought home. But by then Marlatt's argument had gained some traction.

In August 1912, after some modifications, Congress finally passed the Plant Quarantine Act. The act created a board charged with developing an inspection system that it would manage. It wasn't the outright quarantine that Marlatt had wanted, but importation now depended on the exporting country's attention to pests and diseases. If that system was deemed robust, plant imports from those countries were allowed. If not, there would be limitations on the number and use of the imports. Some plants known to carry disease were blacklisted or required mandatory quarantine or fumigation. The act also allowed the USDA to inspect and quarantine imported plants and plants traded between states. Although the act was a positive step, pests and pathogens still slipped through. Some were too small to be seen, or lived in the soil that surrounded the roots, or traveled as eggs under a leaf or as microscopic spores. Frustrated, Marlatt this time suggested the nation ban imported plants altogether. In 1917 Fairchild, who by then had left the USDA, wrote in response to tightening regulations, "We can say to ourselves 'let us be independent of foreign plant production. Let us protect our own by building a wall of quarantine regulations and keep out all the diseases. . . .' But the whole trend of the world is toward greater intercourse . . . less isolation, and a greater mix of the plants and plant products over the face

of the globe." An outright ban was rejected, but in 1918 the USDA
tightened its restrictions by prohibiting the importation of plants with
soil attached to the plant root ball. It also added a requirement that
certain shipments be fumigated. All imported plants would have to
go through Washington, DC, or San Francisco so that they could be
inspected by USDA employees. The change effectively ended the
importation of plants for direct resale to consumers (or plants for
planting versus those used for breeding stock). Whether there would
still be American chestnut forests had those controls been in place a
few decades earlier is an open question.

If you look at a map of chestnut blight's spread from its northern
origins down the spine of the Appalachians, its path looks like ink
spreading through blotter paper. The kill rate for mature trees was
nearly 100 percent. Almost all of the American chestnut trees are
now gone, except for a few odd survivors scattered about the country.
Yet the fungus remains, surviving in the shoots of remnants of once
grand chestnuts that continue to sprout up to this day. A few grow
into mature, nut-producing trees but die young. The blight turned a
once dominant member of the overstory into shrubby understory and
drove a key species into functional extinction.

●●●

AROUND THE WORLD EFFORTS TO CONTROL IMPORTED PLANT DIS-
ease have expanded over the past century. Now there are quarantines
for some plants and routine inspections for pathogens and insects. But
controlling for microscopic fungi hiding in the stem of a seedling or
insect eggs tucked under a leaf even today is a herculean task. In the
1990s a strain of rust fungus called *Austropuccinia psidii* was identified
in Florida, California, and Hawaii. The fungus infects trees belong-
ing to the family Myrtaceae, which includes myrtle trees, allspice,
guava, bay rum, and eucalyptus trees, among thousands of others.
All together there are nearly six thousand species of these evergreen,

aromatic oil–producing trees. Hundreds are known to be susceptible. Unlike many rusts, this one needs no other hosts. It's capable of setting all of its spore types, even the sexual ones, on a single host. The bright, sulfurous yellow spores coat the leaves and buds of infected plants, eventually killing them as the fungus deprives trees of their food. Young trees are the most likely to die. Over the past century *A. psidii* hopscotched around the world, likely traveling in wood or other plant-based products. In the past couple of decades it has picked up its pace. No one knows its origins other than someplace in South America where the fungus is likely endemic.

In 2004, in Australia, some spores were found on eucalyptus wood that had originated in Brazil where there are large eucalyptus plantations (the trees, grown for the paper industry, are not native to South America). As the fungus spread, foresters and others in Australia, where over two thousand susceptible species live, worried about an incursion. Trees and other plants in forests, wetland, street scapes, garden nurseries, and backyard gardens were all at risk. Rust would be catastrophic. A rapid DNA test for the fungus had been developed in anticipation, which is how the spores were identified. The nation developed contingency plans to contain the fungus should it appear again and eventually prohibited imports of wood from countries known to have the fungus. Six years later, in 2010, the pandemic strain was identified by a cut-flower grower in New South Wales. He noticed the fungus growing on his "After Dark" peppermint willows and sent a sample to the local authorities. The finding set off an investigation of the grower's property where a thousand peppermint willows along with other species, including turpentine trees and a bottlebrush tree, were infected. Plants at another property about 5 miles (8.5 kilometers) away also had rust. A couple of months later the fungus was found on even more properties. To contain its spread, 16,000 nursery plants and 5,000 plants growing in the wild were destroyed. The fungus was unstoppable. By December of that year eradication efforts transitioned to disease management. By 2015 it had traveled over 1,200 miles (2,000 kilometers) to northern Cairns, Queensland, and a study published in 2020 concluded that Australia's native guava "is on a steep trajectory towards extinction." In 2017 the fungus was identified in New Zealand where, as in Australia, many of the island nation's trees are susceptible.

A. psidii has gone pandemic, and like blister rust and chestnut blight, it has the potential to cause the extinction of key plant and tree species. But both of those earlier fungi emerged in the first decade of the twentieth century, slipping into the country just as mycologists and disease pathologists were realizing the risks of trade in plants and plant-based items. We know the risk now, and yet rust and other fungi continue to move and be moved. Today, more than ever before, animals and plants are under siege.

Chapter 4

SUSTENANCE

Bananas are a fruit that unites the world. We may not all eat the same variety, but we all know a banana when we see one. Depending on where you live and what kind you eat, they are sweeter or starchier, creamy or tough, all loaded with potassium. Per person in the United States we eat about twelve kilograms (twenty-seven pounds) of bananas a year, more than any other fresh fruit. Elsewhere around the world, bananas are part of the daily diet. After maize, wheat, and rice, they are the world's fourth most important staple crop. In some regions bananas provide 30 to 60 percent of daily calories. Though there are thousands of varieties, most of us in the western world eat only one: the Cavendish. These are the sweet "dessert" bananas we find piled on grocery shelves, hanging in convenience stores, and ever-present in cafeterias. Cavendish are also known as "export bananas" because most are not consumed in the tropics where they are grown but instead are shipped to the United States, Canada, Europe, China, and elsewhere. Of the twenty-two million tons of bananas exported to the United States, Europe, and Asia, most are grown in Latin America and the Caribbean. Many of those are grown in Ecuador, Guatemala, and Costa Rica.

The rest of the world's crop, grown on large and small farms and in backyards around the world, are a variety of cooking bananas or plantains (which are bananas that tend to be starchier and tougher skinned) with names like matoke, Lacatan, Rhino Horn, and Pisang Awak. More than one hundred different plantain cultivars grow in West and Central Africa, where they are a staple food for over seventy million people. For millions of small-scale growers and family farmers the fruit provides both calories and income. An estimated four hundred million people are employed by the industry, picking, packing, and growing the roughly one hundred billion bananas consumed each year. Many of those millions are employed by companies like Chiquita, Fyffes, Dole, and Del Monte, all growing primarily Cavendish. The fruit sustains a $40 billion global industry.

The banana plant is easy to mistake for a tree, but it is the largest known herbaceous flowering plant. Banana plants belong to the genus *Musa* and are recognizable by their large, long, fibrous leaves that umbrella away from the stalk. The plants grow, produce flower and fruit, and die. New banana plants grow as shoots that sprout up from the main stem's base, making each generation a clone of its parent. The cultivated fruits, whether Cavendish or any of the popular varieties of cooking bananas, typically have no seeds. The lack of seeds means that from a genetic standpoint we have been eating the exact same kind of banana for some fifty years.

Depending on how and where it is grown, a typical Cavendish plant bears fruit seven or eight months after planting. The bananas emerge from the plant's weird, oversized, and unmistakably sexual flower, which produces a single bunch of bananas. The large, drooping bunch is composed of dozens of banana "hands," the smaller groups of five or six or more bananas that we find hanging by the checkout counter in the market. By harvest time a full banana bunch weighs between twenty and thirty-five kilos (forty-four to seventy-seven pounds). The plant is an incredibly productive herb.

Bananas are Luis Pocasangre's life work. Pocasangre is the research director and professor at EARTH University in Limón, Costa Rica, where he oversees 439 hectares of banana plants. He grew up in Honduras, the original banana republic, where, he says, "bananas were everywhere and everything." Even the tennis courts on which he learned to play the game were owned by Chiquita. That he decided to devote his career to bananas was the natural course of things. The banana world is incredibly international. Pocasangre received his doctorate in Germany, but before that he studied plant breeding and biotechnology in Costa Rica, while working on a project for a French agricultural organization. Then he worked with Phil Rowe, a legendary scientist and banana breeder. Over three decades Rowe worked for United Fruit in Honduras, where he bred disease-resistant, good-tasting bananas for both export and cooking. Pocasangre now grows several of the hybrids developed by Rowe at EARTH, where he also teaches sustainable agriculture and how to grow bananas to students mainly from rural communities.

Bananas grown for the market need a lot of care, which translates to countless hours of labor. Throughout Pocasangre's orchard the developing bunches are protected inside bright blue plastic bags that protect the fruits from pests. There are plenty of predators who feed on the sweet, starchy fruit: nematodes, thrips, weevils, beetles, bacteria, and fungi, any one of which might scar, rot, or spot the fruit, ruining the perfection we consumers expect. On a conventional plantation the inside of the bags are treated with an insecticide like chlorpyrifos. The chemical is a known neurotoxicant that has been withdrawn from some markets. At least one study of children living near commercial plantations found that they had been exposed to potentially harmful levels of the chemical. For these reasons and others, in 2021 the US Environmental Protection Agency banned its use on food crops. At EARTH the bags are treated with a combination of garlic and onion oil, and an unmistakable sulfurous smell wafts across the plantation.

In addition to the plastic protection, banana workers slip cardboard sheets between each cascading row of banana hands to prevent them from scarring one another. Large, ripe bunches travel from field to processing plant, hanging from a wire tram that runs throughout the plantation like some otherworldly commuters on a trolley. When the banana tram arrives at the processing plant, the fruits are power-washed and examined for blemishes. Workers pick dead "flowers" from the end of every fruit. Then the hands are separated from the bunch and floated in large vats of water as workers pick, pack, and label the best-looking banana hands for export. Each banana you buy has been handled with kid gloves by dozens of workers. Boxed bananas are loaded onto a container truck ready for their journey to the United States, Europe, or elsewhere. Some travel for a week or two before they are unpacked and laid out at Whole Foods or Aldis and labeled "sustainably grown." The rest are sold locally.

BANANA PLANT

One of Pocasangre's research interests is using beneficial microbes and a probiotic sort of fungus called *Trichoderma* to prevent insect pests like microscopic nematode worms. Some strains of this common soil fungus, along with other amendments like composted banana plants, are a biological treatment for the nematodes that eat the plant's roots. Treated plants stand tall. Untreated plants lean on bamboo poles because their roots can no longer support the tall stalks. The alternative control is injection of pesticide gas. The biological treatment works well on bananas, but "bananaeros"—banana growers—are a conservative group resistant to change, so many still rely on conventional pesticides.

EARTH's plantation land is split into widely separated blocks that Pocasangre and others use as a living laboratory to test sustainable solutions for growing the fruits. Between the blocks are forest, wildlife, and river. This arrangement of agricultural crop interspersed with native plants is a form of agroforestry, an alternative to the sweeping, singular monocrops. By *not* planting every inch of soil with crop, pests and pathogens can't easily travel from one host to another. It isn't hard to imagine how the spacing could discourage the spread of a fungus that might otherwise travel from leaf to leaf or root to shoot. Some blocks at EARTH have papaya trees interspersed with the bananas; others combine Cavendish with different cultivars like red Macabu and plantain.

"A real bananara," says Pocasangre, "will be three thousand, six thousand hectares all in the same banana-growing region. No separation. Grown as a monoculture because it's more profitable." Some plantations are even larger. The vast majority of Cavendish are grown in this way, as monocrops, making them ripe targets for fungi.

* * *

A CENTURY AGO A FUNGUS IDENTIFIED AS *Fusarium oxysporum f. sp. cubense* nearly destroyed the banana industry. The disease caused by this fungus (the so-called Race-1 strains that are actually different species) became known as Panama disease or Fusarium wilt of banana. The favored host of the fungus was not the Cavendish but a banana variety called Gros Michel. Gros Michel bananas were the first "big banana." That cultivar's popular history began with its discovery in Southeast Asia. A nineteenth-century French naturalist impressed with the fruit brought a bit of banana plant to the island of Martinique. From there a French botanist brought it to Jamaica. The fruit grew well on the islands, and because they were encased in thick yellow skin, they shipped well too. And it ripened aboard ship. Within decades Gros Michel bananas were popping up on farms all along Central America's Caribbean coast.

By the late nineteenth century, bunches of Gros Michel bananas were offloading at ports in New Jersey, Philadelphia, and Boston. Americans found a new fruit to love. The lucrative combination of desirable and cheap caught the attention of a Cape Cod ship captain and a Boston grocery worker. In 1885 they formed the Boston Fruit Company, the first commercial banana company. Later renamed the United Fruit Company, by 1930 it was worth more than $200 million. The company's dark history and that of other early exporters is detailed in John Soluri's book *Banana Cultures* and Dan Koeppel's *Banana*. As complicated as the business end of bananas was, the agricultural history of the Gros Michel was for a time simple: growers planted it and it grew. By the early twentieth century the tropical fruit was flourishing in Honduras, Costa Rica, Panama, Colombia, Guatemala, and anywhere else growers could profit from planting it. By 1913 Americans were eating on average over twenty pounds of bananas a year per person, and United Fruit had seventy thousand hectares in production.

Of the hundreds of known strains of *Fusarium* most are harmless saprobes living in the soil, sending out filamentous hyphae, and feeding on dead things. But the Race-1 fungi were insidious killers. No one questions how the fungus spread across Gros Michel plantations: the constant growing of monocrops enriched soils with spore. Wherever soil traveled, spores did too: on plants or suckers, on the soles of a worker's shoe, on a truck's tires, with trickles of water, and in floods, hurricanes, and typhoons. Bits of banana plant including leaves, commonly used for packing, could move the fungus farther away.

The industry's response was to cut virgin forest and create new fields. Some fields were flooded and then replanted with shoots from infected plants. Because flooding suffocated both the disease-causing fungus along with much of the beneficial soil microbiome, the disease came back with a vengeance. Old plantations were left to rot. Over the years United Fruit scientists tried and failed to find a suitable replacement or breed a resistant, palatable hybrid banana. Eventually industry losses began draining millions of dollars a year from the bottom line. It wasn't just United Fruit or just Central and South America; the fungus hit growers in Asia, Africa, Latin America, wherever Gros Michel grew. And the growers were to blame. They just kept planting the same old thing in different places.

The outcome was disastrous. It would have destroyed the entire industry except that the Race-1 fungi were also limited *because of* their affinity for the commercial banana. The fungus doesn't infect most other strains of banana, including Cavendish. The Cavendish banana had been known by horticulturists for at least half a century. It was believed to have originated in China, shipped at some point to the island nation of Mauritius controlled by the Dutch, then the French, and then the British (the nation became independent in 1968). When the banana arrived on the island sometime around the turn of the eighteenth century, the British horticulturalist and physician Charles Telfair planted some in his garden. In the late 1820s Telfair sent a sample of the plant back to England, which was planted and then replanted in gardens of wealthy collectors of exotic plants and animals. The banana eventually set roots in the garden of the sixth

Duke of Devonshire, William Cavendish. Throughout the century the Cavendish banana traveled to colonies and locales in the South Pacific, Egypt, and South Africa.

The Americas at that time had no use for the fruit; they had the Gros Michel. Even as the banana industry took off, the Cavendish was considered too delicate compared to the Gros Michel. The banana didn't ship as easily as the Gros Michel, which could be tossed onto ships in bunches. The Cavendish was easily bruised and so had to be boxed. Some thought it wasn't as sweet. Still, it looked and tasted familiar enough, and when Race 1 emerged, it was resistant. Grudgingly the industry switched, changing the process of picking and shipping, and pulling on the kid gloves to provide consumers with the perfect, unblemished banana. Within a few decades Cavendish bananas were growing across enormous monocropped plantations, replacing the Gros Michel. By midcentury United Fruit rebranded with the rollout of the "Chiquita" brand and its eponymous jingle. In 1990 the company rechristened itself as Chiquita Brands International. As a global produce company it remains one of the largest distributors of bananas in the United States.

Now Chiquita, along with the rest of the industry, is facing another round of the dreaded Fusarium wilt. This time it is a highly aggressive fungus called *Fusarium odoratissimum* and known as Tropical Race-4, or TR4. Unlike Race-1 strains, this fungus infects Cavendish plants. To paraphrase Koeppel, the industry could almost have expected this sort of reckoning. And yet they essentially invited the fungus to the table by providing it with a nearly endless monocrop and a means of travel.

●●●

FUSARIUM WILT OF BANANA (WHETHER CAUSED BY RACE-1 STRAINS or TR4) is soilborne. Spores called chlamydospores pepper the soil around infected plants. When a new plant sets roots, the spore germinates, sending delicate hyphal filaments through root and stem. Nutrient and water passageways clog, and eventually the main stem ruptures as fungal hyphae grow through the plant's vasculature. The older leaves yellow, the stem wilts and buckles, while the fungus strangles the plant. Before the plant dies back, the fungus reproduces, sending another generation of spores around the plantation. Like other fungi, *Fusarium* fungi release different kinds of spores. Some, like the microconidia and macroconidia, won't survive long without encountering a host. Chlamydospores, borne from hyphal threads, are resilient. They can last in soil for years (Pocasangre says decades) and explain, in part, how the fungi become entrenched in soil long after the last banana plant has been removed. When banana plants aren't around, the fungus can live in other plants without causing disease—increasing its staying power. With other fungi, some fruit and vegetable crops can be replanted seasons after infection, but after Fusarium wilt bananas cannot. There is no way to remove the fungus other than to eliminate all traces of contaminated soil from the farm or to flood the plantation entirely, depriving the spores of oxygen.

Since its emergence the fungus dubbed TR4 has destroyed millions of hectares of Cavendish crops around the world. A story about the fungus in CNN was headlined, "Why bananas as we know them might go extinct (again)," while *The New Yorker* and the *New York Times* ran headlines that riffed on the old 1920s song "Yes, We Have No Bananas," which was inspired by the first go-round with Fusarium wilt.

First detected on Cavendish plantations in Taiwan in 1967, TR4 may have begun its travels with infected plants imported from Sumatra, Indonesia. A few years later the Taiwanese government issued an emergency order to pull and destroy infected plants and those plants nearby. The job was labor intensive, and many simply pulled or cut the plants and left them to rot on the ground, allowing rain and irrigation water to carry the spores away and providing the fungus with an opportunity to spread. In the following decades it moved across China and then other banana-growing nations. In 2019 it was detected in Colombia. TR4 had arrived in South America.

One way to get ahead of an ongoing pandemic or prevent the next one is to understand how and where the pathogen moves. Gert Kema is a phytopathologist at Wageningen University in the Netherlands; one of his laboratory's projects is tracking TR4's movement. They do so by sequencing fungal DNA, seeking small changes in the genome that they follow through time and space like a trail of breadcrumbs. By the fall of 2021 they had genotyped some two thousand and sequenced around one hundred different strains of TR4 sampled from Asia to Africa and beyond. Despite Kema's experience and large data set, he is reluctant to claim a direct relationship between the strain that occurred in Colombia and strains from other regions. Other scientists suggest that the Colombia strain may have hailed from Indonesia.

Wherever TR4 has spread, Kema is fairly certain that, as it had before, fungal spores traveled with industry equipment or workers walking across contaminated soil, moving diseased soil from one continent to another. Though, he adds, it is possible that despite whatever precautions are taken, TR4 will be difficult to stop. And once the fungus sets into the soil, there is no hope for eradication. What is most frustrating is that when TR4 strikes, it isn't just the crop and the growers who are impacted.

The fungus, says Luis Pocasangre, "is a social problem. It affects everybody. Banks, producers, thousands of field and packing plant workers, scientists, consumers—everybody." Forty thousand in Costa Rica alone are directly employed by the industry, which ripples out to more than one hundred thousand workers in support services. In 2014, several years before the fungus landed in South America, Costa Rica's former agricultural minister told the *Independent*, "Should TR4 invade . . . we would lose $880m of export income. Misery, unemployment, drugs and delinquency would result." "People are in a panic right now," says Pocasangre of the current wave of TR4. Bananas are many things: they are food, big money for industry, and a way to make a living. Sarah Gurr, who studies the impacts of disease on food and commodity crops and the relationship between the two at the University of Exeter, echoes Pocasangre. "In the developing world economies, many countries rely almost exclusively on commodities like coffee and bananas that they can sell and *buy* their calorie crops. Colombia is running on coffee, and Haiti is reliant on banana to buy calorie crops" like wheat, rice, and maize.

The effect of TR4, which is threatening an industry that ships out twenty-two million tonnes of fruit a year, will be felt by all of us to varying degrees. Should the crop collapse, we would lose our favorite breakfast snack, or we might just have to broaden our banana palate and enjoy other more-resistant varieties of the fruit. But millions of others will lose their livelihoods. The story of the fungus encapsulates the problem of a huge global industry wedded to a single clone with no backup plan. The future of our daily banana isn't yet certain, but change is coming.

Fusarium wilt of banana, which impacts mainly export bananas, is worrisome enough, but it isn't the only major fungal disease attacking bananas. If you were to rank diseases that threaten food security, one called Black Sigatoka would rise to the top. The fungus that causes it, *Pseudocercospora fijiensis,* was first noticed in the 1960s in Southeast Asia. Unlike in Fusarium wilt, the spores travel by air, which means a spore blown by the wind can travel from one farm to another and there is little a farmer can do to keep it from landing on her own crop. Within hours of landing on a banana leaf, if there is enough moisture, the spore will germinate. Like other fungi, hyphae in search of food invade plant tissues through the stomata. From there the fungus weaves through and around the plant leaf's cells before emerging again through the stomata. When the leaves die, the plant, drained of resources, has less energy to put into fruits, and the plant becomes less productive. Even dead and

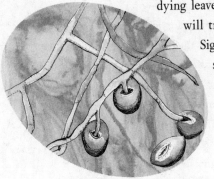

HYPHAE INVADE THE LEAF'S STOMATA

dying leaves release millions of spores. Some will travel over a hundred miles. Black Sigatoka is not only prolific; it is also sexual, some would even say "supersexual." As it churns out tens of thousands of spores, the potential for genetic diversity is huge. And unlike Fusarium wilt, Black Sigatoka infects both the dessert and the cooking bananas.

If there is one arguable saving grace, it is that Black Sigatoka is manageable with pruning and fungicide. Pruning away infected leaves physically stops spores from spreading. But on large plantations or where the fungus flourishes, it isn't enough. In some locations where the fungus grows best, planes drop fungicide on the plants every five days or so; some crops rely on sixty or more applications to make it through the growing season. Many subsistence growers don't use fungicides because they are inaccessible or too expensive, which makes controlling the fungus tricky. When Sigatoka comes, if growers aren't spraying or pruning, they can lose between 30 and 80 percent of their yield. But small growers are also doing something that the big Cavendish plantations are not. They are growing many different varieties of banana and practicing agroforestry. Because the fungus needs some sunlight, when bananas grow as an understory crop in the shade of taller trees as they would in the wild, Black Sigatoka can't thrive.

The banana industry has a history with a relative of Black Sigatoka. Beginning in the 1910s, Yellow Sigatoka, a disease caused by *Pseudocercospora musicola* and named for the yellowing spots on leaves, began spreading through the banana lands. Unlike Panama disease, this one was more manageable. Infected plants were doused with the go-to fungicide of the time, the copper-based Bordeaux mix. Because the fungus lived on the leaves and was easily accessible and because the mix would stick to the leaves, spray treatments were a remarkable success. But the work on such a large scale was messy and labor intensive. Workers hauled the copper mix around the sprawling fields and sprayed hundreds of gallons onto plants by hand dozens of times a season. After harvesting, the copper residue was removed by dunking bunches of bananas into large vats of acid followed by a rinse of water. While consumers in the United States and elsewhere enjoyed their bananas, workers suffered. Recounting the personal impacts of Sigatoka control, labor activist and University of Washington historian Steve Marquardt wrote that copper sulfate built up on workers' skin and clothing "until it formed a virulent blue-green crust." Even months later, in workers who were no longer exposed, wives and other family members said that "the mucous membranes of former *pericos* remained greenish, and they still expelled green-tinged sweat." *Perico* means parakeet; workers used the term because their skin and clothing took on an indelible copper green tint. A condition dubbed "sprayer's lung," which could be chronic and deadly, with symptoms mimicking those of tuberculosis, was also common.

In 1942 United Fruit Company workers wrote to Costa Rica's president Rafael Ángel Calderón Guardia: "We spray workers, based on the bitter experience of our work, tell you that headaches, night coughs, and bad eyes are all common among us, that is, we suffer in our vision, our brains, and our lungs; we are very prone to tuberculosis." But the use of spray workers continued throughout the 1950s; only in 1958 did the practice of spraying of the Bordeaux mix end. The hundreds of kilos of copper that had been sprayed in Costa Rica over the years left a lasting impact on the soil, rendering some of the most heavily treated land useless for planting.

After World War II new chemicals became available to growers and consumers alike, such as the chlorinated insecticides DDT and toxaphene. New fungicides like mancozeb and benomyl that were easier to apply and often more toxic than copper also came onto the market. Mancozeb was one of the first alternatives to become available in the banana industry. It kills fungi by disrupting enzymes involved in key metabolic functions and works on contact. The benomyl-based fungicides, available in the late 1960s, interfere with cell division. Both fungicides are toxic to humans and wildlife and have been linked to birth defects; mancozeb may also interfere with the nervous system. And only a decade after using these new fungicides, some targeted fungi were showing signs of resistance. In the 1980s a new class of fungicide, the azoles or triazoles, became popular. These chemicals inhibit an enzyme needed by fungi to build their cell membrane. This mechanism of killing fungi is shared with the azole antifungal drugs commonly used to treat fungal infections in humans. Recently, use of these fungicides in some agricultural settings has been linked to drug-resistant human fungal pathogens.

Though pesticide use and application conditions are safer today, heavy pesticide use remains a problem for both banana workers and for those who live around large commercial plantations. The constant pressure of fungicide application combined with a prolific and genetically diverse fungal population is a recipe for resistance and in turn more fungicide. The more we use, the more we select for microbes that evolve to survive the chemicals we use. This is the chemical treadmill growers are on, and they can never quite catch up.

• ◉ •

IN 2015 LLOYD'S OF LONDON, THE PROMINENT INSURANCE MARKET, imagined a "food shock" scenario, a situation in which production of the world's top four crops decline. Lloyd's envisioned floods and drought, both exacerbated by climate change, in addition to massive fungal pandemics, including stem rust and another rust that attacks soy. The scenario imagined the worst; the instability of food crops could bring terrorism, political instability, food riots, and, as they wrote, a cascade of "economic, political and social impacts." It is a typical dystopian movie script, but the impact of fungus on crops isn't just science fiction and does have the potential to impact us all. Rice feeds half the world, and wheat is grown over more land area than any other agricultural crop. The grain provides more than four and a half billion people with around 20 percent of their calories and protein. Wheat, rice, and maize are all significant staple crops, and all are threatened by one fungus or another.

Ever since the dawn of agriculture, wheat has been chased by a rust fungus. Over ten thousand years later we are still dogged by rust, as an ever-growing population relies on wheat. In 1999 a highly virulent strain of rust called Ug99 was discovered on wheat in Africa. Many feared it would spread around the globe. Gurr says that the Ug99 strain is "a tiny part of a huge story. There are many thousands of wheat stem rust strains." And there are other fungal diseases of wheat, one of which likely jumped into wheat from a disease called rice blast sometime in the 1980s. Gurr and her colleagues speculate that the practice of growing large monocrops of rice provided the fungus with an opportunity to evolve and infect a new host: wheat. Yet despite the problems with monocultures, Gurr doesn't see an easy way out. "We can pretend there is hope," she says. But if we are going to feed the world's growing population, "monoculture is the only way to grow more." This means fungal pandemics will continue to threaten our food, from wheat to rice and maize to bananas—and beyond.

Chapter 5

NIGHT

For at least half a century and probably longer, bat watching had been a summer tradition in my town of Montague, Massachusetts. The most common bat is the little brown bat, or *Myotis lucifugus* (*Myotis* means mouse-eared; *lucifugus* means to flee the light). These small bats have glossy fur and rounded ears. Just around twilight neighbors would wander down to the old Congregational church and peer up toward the second-floor attic. Inevitably the bats would appear, slowly at first. A single bat would drop from the eaves, falling into the night sky before catching air—then another, followed by two or three. Moments later they would pour from the structure into the dusk. Hundreds of little brown bats leaving their roost, taking flight, dipping, rising, and turning on a dime in search of moths, beetles, and other insects—nocturnal hunters of an airy ecosystem. The church was a place for pregnant bats to congregate, give birth, and then nurture their young. In the spring females would arrive at their roost pregnant, having stored sperm since the mating season in the fall, before hibernation.

When the air warms and the bugs come out, mother bats go on the hunt, the newborn nursing pups sometimes clinging to them as they fly. Little browns consume about one-third to one-half of their body weight each day. Hundreds, sometimes thousands of moths and beetles will be caught and eaten, enough to provide the small mammal with a few fractions of an ounce of insects. She will digest their bodies, often tearing off wings and other parts before consuming. The nutritious bits of her catch are stowed away as fat or turned into milk for her pup, while the rest she excretes as guano, or bat poop. Scale up the nightly feeding frenzy from a single night to a whole season of bats wheeling overhead, and those few grams of moths, beetles, and mosquitoes add up to several hundreds, possibly thousands, of metric tons a year consumed by little browns across the continent.

When the air chills and bugs disappear, the bats seek out a cave or mine where winter temperatures are moderated, predators are few, and there is a bit of water; these are their hibernacula. One of the largest in the Northeast is Aeolus Cave in East Dorset, Vermont, which sits slopeside in the Taconic Mountains. Bat scientists keep tabs on hibernating bat species like the little browns by entering hibernacula once each winter and counting them, careful to avoid disturbing their subjects. For decades, tens or hundreds of thousands of bats are estimated to have settled throughout Aeolus each winter. The cave is about 70 miles (112 kilometers) from Montague, and it is likely that at least some of those Montague church bats overwintered there, huddled wing to wing with thousands of others.

A little over a decade ago the bats began to disappear. There was a rumor that the church in Montague cleared the attic and blocked their passage. But that wasn't the case. One by one, then in whole groups, bats were killed by an innocuous-sounding fungus called white nose syndrome, named for the puff of white visible on their mouse-like snouts. The disease, caused by the fungus *Pseudogymnoascus destructans* (Pd), is killing bats across North America.

Killer fungal diseases are rare in mammals. Of the millions of known fungal species, only a few dozen or so are known to tolerate warm bodies like bats' or ours. Most mammals simply run too hot. We know that humans have become increasingly susceptible to infection because of modern medical interventions that can suppress immunity; some fungi manage to survive our warm bodies. Still, the number of disease-causing fungal pathogens that tolerate the heat remains small. We, and many other mammals, maintain our high body temperatures all year round, but many bat species do not. Little brown bats hibernate, and when they do, their body temperatures drop. It is precisely this sort of life-history loophole that white nose syndrome exploits.

Since 2007 the syndrome has killed millions of bats—including little browns, big browns, northern long-eared, tricolored, and the already endangered Indiana bats. All are hibernating species. The epidemic has emptied bat caves, belfries, attics, and barns across the United States and into Canada. The losses are sobering. Pre–white nose populations of little browns in the East numbered in the hundreds of thousands; they are now down to tens of thousands at best. They, along with tricolored bats and northern long-eared bats, have lost 90 percent of their population to white nose syndrome. In 2015 northern long-eared bats were listed under the Endangered Species Act as threatened. For some of those populations the losses are unsustainable. We are losing our bats to a fungus.

LITTLE BROWN BAT / LONG-EARED BAT

• • •

ONE-QUARTER OF ALL KNOWN MAMMALS ARE BATS, OR CHIROPTERA—
hand-wing animals. Bats are most unusual among mammals for their
flight. Some sixty million years ago, evolution began shaping bat wings
from finger bones and skin. If you stretch out a bat wing, each long,
delicate bone and each joint corresponds with the bones and joints in
your hand. A membrane called the chiropatagium stretches from the
thumb between each of the bat's fingers. Another membrane fills the
space between the hind legs and tail. These modifications endow bats
with the gift of flight, and the combination of membrane and long,
slender fingers provides bats with remarkable maneuverability. Bat
flight in slow motion mesmerizes as wings stretch and fold, stretch
and fold. If you watch bats hunting at night, you may have noticed
how they dart and turn, chasing down insect prey, or heard the quiet
swoosh as they skim overhead. Most remarkable is the bat's reliance
on echolocation to navigate and hunt. Much like the pinging sonar
used by ships and dolphins, they send out sounds we cannot hear,
recapturing them as they bounce off prey and other objects.

Little browns echolocate to find insects and to navigate across
rooftops, telephone wires, and fields and through the trees. Their
ears are an essential organ that, like other vertebrate ears, are a Rube
Goldberg–like wonder of funnels, gongs, and electrical signals. When
we talk, or a dog barks, or a bat calls out, vibrations travel through
the air as waves of pressure and interact with the hairs, bones, skin,
and nerves in the ear. As sound enters the outer ears, or pinna, it is
funneled toward the tympanic membrane, or eardrum, of the middle ear. Three small bones of the middle ear—the incus, stapes, and
malleus—help moderate and transmit incoming sound vibrations to
the inner ear, the snail-shaped organ that transmits sound to the brain.
Most of us hear waves with frequencies ranging from 20 hertz to
20,000 hertz. If we are listening to music, violins and piccolos emit

sound waves up to 3,500 to 5,000 hertz, while the tuba pushes air at a much lower frequency ranging from around 40 to 375 hertz. Our range is good enough to enjoy an orchestra but limited compared with other species like cats, dogs, beluga whales, and bats, whose ears can hear in the 100,000 hertz range. Little browns emit and receive sounds between 40,000 and 80,000 hertz. Big browns echolocate around 25,000 to 65,000. When the sounds are slowed into our hearing range, bats sound like chirping birds. Both brown bat species send out sounds that can be as loud as 110 decibels. If we were able, we would hear the noise equivalent of a symphony of jackhammers. It would be enough to cause hearing damage and sonic confusion, even for a bat. But within the bat's ears is a small muscle that clamps down their tiny ear bones, clicking them open and closed dozens of times a second and protecting them from their own sound.

The majority of bat species feed on insects, fruit, and nectar. Some eat meat: fish, frogs, small rodents, birds, and other bats. The few true vampire bats feed on the blood of easy-access animals like chickens, dozing pigs, and cattle. At the tip of a vampire bat's snout is a thermosensor, which helps it locate blood flowing under its host's skin. Both the genome and microbiome of vampire bats have adapted to this eating habit, enabling them to survive the pitfalls of living on a liquid diet, which likely includes blood-borne microbes and toxic levels of iron.

Central and South America are home to hundreds of bat species, including the vampires. The smallest known bat, the bumblebee bat, fits on the tip of a finger and lives in Thailand. One of the world's larger bats is the fruit-eating giant golden-crowned flying fox, with a wing span of five to six feet. It is listed as endangered along with more than fifty other bat species, most of which were declining well before white nose syndrome.

Of the forty-seven different North American bat species, about half are migratory. The rest ride out the winter months hibernating in caves and old mines, roosting in churches and attics or trees during the spring and summer days. The silver-haired bat, named for its white-tipped fur, roosts in tree cavities in the summer but migrates south for the winter. They can be found from Alaska to Mexico. The northern long-eared bat (which is a mouse-eared bat like the little brown) is a tree-rooster and squeezes under a strip of bark or into a crevice or cavity during the warm months and then hibernates in the winter. The big browns may hang with the small browns in an attic or barn roost. All of these, along with hundreds of other species, are vesper, or evening, bats, so called for their nocturnal nature. There are several other groupings with descriptive names like leaf-nosed, free-tailed, ghost-faced, and sucker-footed. When most of us see them, we think "bat." But they are far odder looking and more interesting than we can imagine.

● ● ●

IN THE EARLY 2000S JON REICHARD WAS WORKING ON HIS PHD AS A graduate student of Thomas Kunz, a world-renowned scientist at Boston University. Known as "Bat Man," Kunz was a tireless, innovative, and broad-ranging researcher. And he was all about bats: physiology, ecology, conservation, and aeroecology, a field he helped develop to better understand how aerial life interacted with the physical nature of the oddly fluid environment in which bats lived. His work and that of his dozens of students provided real-world insights into animals that were not well understood and difficult to study *in situ*. Reichard studied Mexican free-tailed bats (*Tadarida brasiliensis*) in Texas. These charismatic creatures, which some say are the fastest mammals on the planet, roost by the million under the Congress Avenue Bridge in Austin, Texas. It is the largest urban gathering of bats in the country and attracts tourists from around the world. Kayakers, canoeists, tour boats, and pedestrians gather at dusk to watch the living stream of bats as it flows from the bridge and along Lady Bird Lake. Reichard measured temperature signals and heat loss in bats that lived in Texas Hill Country caves. Because it is difficult to do this without disturbing them, he used a thermal sensing camera, a new device at the time. He had thought after grad school he would explore how bats managed to prevent overheating on long flights, particularly those like the free-tailed bats that migrate. Or, he thought, maybe he would study north-eastern bats and whether they eat enough of the right kind of insect to protect local crops like apples or cranberries.

Reichard was finishing his dissertation in the winter of 2007 when Kunz's lab got a call. Something odd was happening with the bats in a cave in New York: could they come take a look and bring that thermal camera Reichard had used in Texas? The previous winter a caver exploring a cave system called Howe Caverns near Albany, New York, noticed bats flying around during the daytime in the winter, which was unusual. A few months later there were dead bats in three caves and mines nearby, many dotted with a little white, cottony fungus. The cause of death was a mystery; fungus wasn't known to kill bats. The finding was odd and disturbing, but many hoped it was a one-off, something that affected just a few caves and may have been catastrophic for those populations but would remain a local-ized catastrophe. The next winter biologists entered a different set of caves and mines south of Albany. Dead bats littered the floor. This time they were shocked. The live bats remained in a state of torpor, a typical condition during hibernation where physiological activity is slowed, but that was odd, too. Surveys often disturb the bats, briefly rousing them. These bats didn't move. Shaken, the scientists called in other experts including Kunz and Reichard. Maybe body tem-perature would provide some clues to the unusual sluggishness, so Reichard set off with his camera. What he saw was terrifying. The very bats he had imagined building a research program around were dead or dying.

The following winter Reichard visited the Aeolus Cave in Vermont. "The floor was littered with thousands of bat carcasses. There were dead bats hanging with sick bats clustered up next to them on the wall. There were ice formations in the cave with bats frozen within; bats were crawling around in snow presumably wanting water; and in the mouth of one cave, a tufted titmouse was scavenging bats." Other scientists described caves that smelled like death, finding mice eating moribund bats that were too ill to fend them off. When the disease first hit, the massive die-offs shocked and saddened biologists. The maternity colonies around New England that Kunz had studied for decades were suddenly silent and empty. Reichard and a generation of bat scientists were now on a mission they never could have imagined: saving a once highly abundant species or documenting its extinction.

The normal body temperature of little brown bats isn't all that different from ours, ranging from 35°C to around 38°C (95°F to 100°F). During torpor their body temperature cools to between 4°C and 10°C (39.2°F and 50°F) nearly matching that of the cave. Metabolism slows to a pace just enough to keep life-support systems running; heart rates normally in the hundreds of beats per minute might slow to a couple dozen beats; breathing slows to barely a single breath per hour. Torpor helps bats conserve energy through the winter. But every ten to twenty days or so, hibernating little browns rouse. Their bodies warm up. They may fly around a little, sip some water, or relieve themselves. They might even catch some sleep (torpor doesn't provide the same respite as a good deep sleep). Each time bats rouse, they burn up a bit of their fat reserve. This is normal. Over millennia, bats have evolved a seasonal routine of summer fattening and winter torpor, gaining just enough weight, about two grams on average, to survive the winter. This adaptation provides a physiological and behavioral balance enabling the species to survive cold winters with little or no food. It is a tough way to live, but compared with other small mammals some bats have a relatively long life span. On average little brown bats live six or seven years, with some elderly bats living thirty or more. At least they did before the emergence of the fungus.

White nose syndrome is caused by a psychrophilic fungus; it thrives when temperatures drop to between 4°C and 20°C (39.2°F and 68°F), which is cold even for fungus. It flourishes in and on bats in torpor. The bats most likely become infected when they return to their cave for the winter. They may pick up a spore from the cave floor while drinking from a puddle, or from the cave wall, or when they nudge up against their neighbors before sinking into torpor. The spores released by Pd are called conidia and resemble microscopic caraway seeds. The fungus feeds on keratin, and skin is rich in keratin; when the spore germinates, it releases enzymes that dissolve skin. Under the microscope colonized wings appear covered in a tangle of hyphae that wind around hair shafts like a vining weed choking a garden flower. Lesions dot the wings until some resemble a thoroughly moth-eaten sweater. The damage disturbs the bat's electrolyte balance and causes dehydration. Infected bats may also move in and out of torpor more often. Hyphae growing on the skin will grow additional internal walls called septae that enable fragmentation of the hyphae into conidial spores, which then break away like so many bits of toilet paper from a roll. Both spores and hyphae can spread from bat to bat. When the bats leave the cave, the spores and hyphae remain on the walls and floor, surviving the spring and summer months until the bats return. No one knows how long the spores can manage without a host; scientists estimate years. Relatives of the fungus are saprobes that feed on the dead or dying. Their hyphae may survive by scavenging on bits of shed skin and hair or dead insects scattered about the cave floor.

For reasons scientists have yet to fully understand, infected bats rouse from torpor more often and so burn more fuel during the winter. This leaves them weakened when it comes time to actively hunt and bear pups. In the spring some survivors may leave the cave but then die later. Others may simply lack the energy and body conditioning needed to bear pups.

Torpor provided an opportunity for the fungus, but bats are not entirely defenseless against invasive disease. Bats have a well-developed if not idiosyncratic immune system. Like other mammals, they benefit from a robust and redundant system that responds rapidly and nonspecifically while at the same time prepares to protect against future infections. *Unlike* other mammals, bats are notorious for tolerating viruses, a characteristic that makes them unique potential vectors for viruses, including the coronaviruses that cause SARS and COVID-19. Why this is the case and how it might impact the bat's response to other pathogens isn't fully understood. When a disease-causing fungus infects the skin, it sets off a cascade of events as immune cells congregate around the invader. But how the system reacts when bats hibernate and are in torpor is unclear. Early studies with other hibernating animals suggest that the activity of cells that tend to be the first responders are at best tamped down. Another study has shown that the B-cells and T-cells essential for producing antibodies may be sequestered away rather than circulating in the blood. Marianne Moore, an ecological immunologist who also studied with Kunz, suggests that if you were to take a blood sample from a bat during torpor, you would be hard pressed to find any immune cells, whereas in the active season there would normally be hundreds of such cells. Parts of the immune response are significantly shut down during torpor. There are some indications that bats' intermittent arousals give them some ability to mount an inflammatory response, but against white nose syndrome it apparently isn't enough. Hibernators probably depress the immune system during torpor to save energy, which Moore points out, makes sense. Under normal conditions, a hibernating colony—like a giant quarantine pod—has little chance of exposure to any new pathogens. And most mammalian invasive diseases require a warm body, so torpid animals usually aren't good hosts. When the bat arouses, the immune system revives. If the wings are full of fungus, as they are with white nose syndrome, this sudden activity may present another problem.

Years ago researchers suggested that some of the worst wing damage might occur days or weeks after the bats have aroused from hibernation when body temperatures soar and the immune system resurrects. They documented bats with wings so ragged that it would be hard to fly, let alone hunt. But the bats were sometimes found far from their hibernacula. The scientists hypothesized that perhaps the damage happened not in the cave but later, after the immune system kicked in, as a sort of rebound response. This phenomenon was similar to what had been observed decades earlier in HIV patients who exhibited a condition called immune reconstitution inflammation syndrome, or IRIS. Stuart Levitz, who worked with AIDS patients early in his career and witnessed the rise of antiretroviral drugs, is familiar with the syndrome. AIDS patients infected with *Cryptococcus*, he says, can have huge numbers of fungal organisms in the brain, "millions of organisms per gram of cerebrospinal fluid," and yet a patient may be asymptomatic. But when antiretroviral treatment begins and the immune system recovers, patients can experience an immunologic storm—a shock to the system. The fungus, he concludes, wasn't causing damage (although if left untreated, it would eventually cause symptoms and kill the patient); it was happily growing in the brain. The damage happens when the immune cells reconstitute. In humans, now that the destructive sequence is understood, patients are cleared of the infection first and *then* treated with the antiretrovirals, which will enable their immune system to rebound. Infected bats don't have that sort of option.

The little browns, once the most common bat in the eastern part of the continent, are now listed as endangered by the Canadian government and the International Union for Conservation of Nature. (The United States has yet to make this call.) The bat count in Aeolus is severely reduced. Some little brown populations are functionally extinct—meaning that like the chestnut trees, there aren't enough breeding pairs for the population to survive. Such a large loss is bound to have ecological consequences. All of the insects consumed by bats could instead consume a farmer's crop. Kunz and others once estimated that bats are worth on average about $23 billion a year to the US agricultural industry. That figure doesn't include their role in the health of forests and nonagricultural systems. Bats that eat fruit help disperse seeds, and those that feed on nectar are pollinators of desert and tropical flowers like agave and saguaro, mango, banana, and guava. Simply put, bats are underappreciated contributors to environmental stability. Some of us may not love them, but depending on where and how we live, our world would be deficient without these enigmatic and weird winged mammals.

．●．

HUMANS HAVE ALWAYS BEEN ON THE MOVE—TRADING, FLEEING, conquering, farming, exploring. But today's pace is unprecedented. In the past century, by the kilometer per hour, we have increased our travel roughly a thousandfold from the generation before us—not counting space travel. Faster ships and larger and more efficient airplanes are moving people in unprecedented numbers and rates. In 2019 nearly 80 million international visitors arrived in the United States; half of them crossed an ocean or two to get here. Domestically, we make some 2.3 billion trips in a single year. We travel for business and leisure and to improve our lives. The sheer number and the pace of humans moving about the planet is mind-boggling, as if there is some giant conveyor belt constantly moving humanity around the globe. We hop off in Paris or San Francisco, Hong Kong or Accra, and then move along. In the late 1970s a fungus infecting wheat called yellow stem rust is believed to have traveled from Europe to Australia, clinging to some traveler's clothing. Scientists think that Pd made its

trip across the Atlantic in a similar way—picked up on boots, clothing, or a backpack in Europe and deposited in Howes Cave in New
York. The cave is connected to Howe Caverns, an underground system where the slow drip of limestone and water formed statuesque
stalactites and stalagmites. Every year some two hundred thousand
visitors from all over descend the 150 feet into the cool dark of the
cave to see the natural wonder.

Jeff Foster is a disease ecologist at Northern Arizona University
who has sought to find the origins of white nose syndrome. One of
his first tasks was to develop a genetic test to identify the fungus. It's
tricky, he says, because you have to be able to distinguish the fungus that causes white nose syndrome from its close relatives because
they too live on cave floors and feed on hair, guano, and other bits
of organic material. Foster uses a rapid DNA test to identify genetic
material specific to the Pd fungus. Like others who track disease, Foster looks for small changes in the DNA that occur naturally over time.
But because the disease introduction was so recent, there are fewer
obvious changes. So Foster needs more genetic material with more
genetic markers. Despite all the technological advances over the years
and the relative ease of DNA analysis, tracing the route of the fungus
to Howes Cave has been an exercise in frustration. "To know *exactly*
where it came from, you need to have sampled where it came from.
And it's clear that we never have. And no one else has either." But they
do have what appears to be close relatives. The fungus appears to hail
from somewhere in Europe, possibly Central Europe, a place where
white nose syndrome disease is endemic.

In 2013 Foster's colleagues at the Smithsonian started sampling
DNA from bats collected in the nineteenth and twentieth centuries
from North America, Europe, and East Asia. One of those bats was a
medium-sized, long-eared woodland bat called a Bechstein's bat. The
creature had been caught on May 9, 1918, by a collector in the Centre-
Val de Loire, France, and then dried. At some point it was shipped to

the United States where it sat on a shelf in the National Museum of Natural History for nearly a century. When sampled for fungal DNA, it turned up positive for Pd. None of the nineteenth- or early twentieth-century bats from North America tested positive. Here was one bit of evidence suggesting that the fungus was present at least a century ago in Europe but not in the United States. Most likely the fungus has been in Europe (or Eurasia) for thousands of years, and now European bats live with it despite being infected. Reichard adds that some of those bats may weather the fungus better than others. Indeed, for European bats the fungus has become a background infection. This also means that Pd fungal hyphae and spores are likely to contaminate the floors of European bat caves shed from mildly ill or asymptomatic bats.

When the fungus causing Pd traveled across the Atlantic, it likely traveled as a spore in a clump of mud or on the clothing of a cave-loving tourist. As with other kinds of infectious diseases, success depends on the number of infectious particles combined with some number of introductions. Many are likely to fail. If fungal spores don't find a hibernating bat's wing on which to germinate, it is not hard to imagine how they might become buried in mud on the cave floor or be washed away, or eventually decompose. But at least once, a spore landed on a bat and began to grow. Most bat scientists including Reichard agree that this is the most likely scenario.

Foster tracked Pd across the United States, where it spread rapidly with few genetic changes. The results weren't all that unexpected; bats rather than humans appear to be the primary vectors for now. But then it appeared in Washington State, which was odd because most bat populations don't really cross the Great Plains. The grassy flatlands that stretch from west of the Mississippi to just east of the Rockies are like "the great divider of North America," Foster explains, and many animals species don't often cross. How the fungus made the jump isn't yet clear, but cross it did—with or without human assistance.

White nose syndrome is here to stay, leaving scientists to wonder for those bats that do survive how the fungus will impact their populations and their physiology. At some point in history, it was probably new to European bats too, and now they live with the fungus. Most likely the bats evolved ways to survive. Recently, scientists have seen some hopeful signs in at least some US populations: reports of bats scattered across the Northeast that seemed to be surviving the fungus.

In 2018, Reichard, along with several other scientists including lead Tina Cheng who was working with the nonprofit organization Bat Conservation International, reported on six populations of little browns living in the Northeast, including those remaining in Aeolus Cave. Because they had studied these sites in both 2009 and 2016, they had the rare opportunity to compare bat populations from before and after white nose syndrome hit. They knew that infected bats arise more often from torpor and that arousal burns up energy. The bats, which normally survive on only a couple of grams of fat during winter, were already living precariously. Was it possible that bats with more body fat going into hibernation had an edge on survival? The study found that surviving bats at some sites were significantly fatter in 2016 than in 2009. Putting on a few extra grams, it turned out, seemed to be a good thing, aiding in the bats' survival. The cause of the gains whether genetic or environmental—maybe there were just that many more insects to catch—would have to wait for another study. Still, the conclusion suggested some strategy for conservation efforts, such as helping bats fatten up before the cold sets in.

These weren't the only survivors. Generally, depending on the location, less than 10 percent of little brown, northern long-eared, and tricolored bat populations hit by white nose syndrome have managed to survive. The observations provide some hope that bat populations might evolve to either tolerate or resist disease. If they can, it would be a remarkable feat in the course of just two decades. But, then again, this sort of rapid or contemporary evolution, if that is what it is, isn't unheard of in the animal world.

• • •

IN 1973 A PAIR OF YOUNG SCIENTISTS ALONG WITH THEIR TWO daughters and a few others motored over to a remote island in the Galápagos. They grabbed their gear and leaped onto a tiny island called Daphne Major, the tip of a volcano poking up from the water. Their intention was to study life on the inhospitable site for a couple of years. Instead, the scientists, Peter and Rosemary Grant, returned year after year for forty years. The couple's time on the island would come to transform our understanding of evolutionary processes.

Both Grants are evolutionary biologists by training; Peter's strength was ecology while Rosemary's was genetics. When they first jumped onto the rock in the South Pacific, their goal was to study the island's finches for clues about how new species of birds evolve. These were the finches made famous by Charles Darwin; the differences in beak size and shape contributed to his understanding of evolutionary processes. Of the thirteen species of finch he identified, Darwin hypothesized that all descended from a single species. Their beaks and eventually the species had been shaped by environment and circumstances like island geography, which naturally separated populations. Over thousands of years through the process of natural selection the beaks had evolved to suit the environment. The fruit eaters had stubby beaks while insect eaters developed more slender beaks. Others had strong beaks able to crush the nutrients from tough seeds. The Grants studied the medium ground finch (*Geospiza fortis*), a seed eater. They measured the birds from tip to tail, studied their beaks and the food they ate, and took note of rainfall and sunny days. When the sun was out, the island was hot and dry. When it rained, everything was wet for days on end. The two years turned into several, and in 1977 something peculiar happened. It didn't rain for over a year. The drought conditions altered the available food. Two main sources for the medium finch were big seeds and small seeds. Like all populations of animals, there was natural variation among the medium finches. Some had bigger beaks and others had smaller ones. Bigger-beaked birds tended to eat the larger seeds, while those with smaller beaks ate the smaller seeds. As the drought progressed, the smaller seeds became scarce. A student of the Grants who had been working on the island at the time wrote to the scientists that the birds were dying. This was not the making of a good field season. When the Grants arrived on the island that year, they began measuring birds and beaks. Those with small beaks had largely died out, while those with beaks able to crack larger seeds survived. This, they recalled in a later interview, was the first glimmer. The offspring of these survivors were large beaked. The population had shifted toward larger-beaked birds, and the shift was heritable—genetic. When weather conditions shifted, favoring smaller seeds, so too did the population. In that few years' window of disastrous opportunity, the Grants witnessed the machinations of evolution.

Evolution by definition is a shift in gene frequency within a population, and natural selection is one force through which this happens. When the Grants identified finches with bigger or smaller beaks and saw that these changes were heritable, they suggested a shift in certain genes had occurred. Natural selection acts on genetic variation within a population. Sometimes that variation emerges from relatively new mutations—as may happen for some antibiotic-resistant bacteria and fungi. But selection also acts on existing variation—genes, or traits, scattered throughout the population.

In this case ancestral genes present in the population rescued the birds from a somewhat random but not unprecedented change in climate. Genes like this may be selected now and again, over and over through the years, so that they are maintained in the population as a whole. They are like the unused sweater in the back of the closet, useful only for the rarest of cold winters. What makes the Grants' finding even more extraordinary is that the finch lays just three or four eggs at a time, which distinguishes them from rapidly evolving insects and rodents—animal species that are known to evolve rapidly in response to pesticides or rodenticides aided by enormous numbers of offspring. Though the Grants' work didn't focus on disease, their discoveries give some hope to scientists like Reichard and others who have watched populations of bats, frogs, salamanders, and others challenged by pandemic diseases. Perhaps these animals have some useful genes in the back of the closet.

It makes sense that populations living under constantly changing environmental conditions—dry spells, big seeds, small seeds, a flare-up of local disease—retain some genetic variation enabling them to cope with change. This so-called standing genetic variation provides a population with an escape hatch. Most living things are alive for a blink of time, but populations stretch over eons, collectively, sequentially living through natural environmental change. Now life is faced with a new challenge, that is, *unnatural* changes to the environment: the diseases and other invasive species we as humans move around the globe, the land we clear, and the climate we have changed. As a result we are witness to and the cause of the planet's Sixth Mass Extinction.

For North American bats white nose syndrome may be something new, but the risk of starving because they can't catch enough food, or an oddly cold or warm winter that might cause them to use more or less energy, or even exposure to other diseases likely is not. Sometimes there are advantages for species that have survived disparate environmental challenges.

· ● ·

WHILE WHAT SOME CALLED THE "FAT BAT" STUDY WAS UNDER WAY, graduate student Giorgia Auteri and her adviser Lacey Knowles wondered if the bats' survival could be explained through genetics. What if previously unimportant variation in some characteristics had, when bats were faced with Pd, suddenly become important? One example would be the ability to conserve energy when there isn't much food. If variation in this characteristic was based on genetics, it could give a selective advantage to bats facing the novel challenge of surviving Pd. Possible variations could be slightly thicker skin, different metabolic rates, or changes in hibernation patterns. If a favorable genetic makeup was available, bats, or at least some bat populations, might survive white nose syndrome through the process of evolution. Auteri and Knowles studied bat populations in northern Michigan, where the fungus was first detected in 2014.

Auteri is old enough to recall the days before white nose syndrome and watching as it made its way west across the country. When she began her studies, it wasn't yet in Michigan. Two years later it was. She's also young enough to be part of a cohort of people who live in a world where white nose syndrome simply exists: it is the way the world is.

Auteri began working with bats as an intern out of college, doing surveys of bats in the Great Smoky Mountains. Her next job was monitoring dead bats under wind turbines and noting any endangered species that had flown into the turbines. "More bats than birds are killed by wind turbines. My first job was 'how many cool different bats!' and my second was 'kind of sad!'" Migrating bats that are less susceptible to white nose syndrome are most affected by the turbines, and those that hibernate are more likely to die from white nose syndrome than from turbines.

At the same time, a growing body of research had demonstrated that the rapid evolution observed by the Grants and others wasn't all that unusual. Other animals, including vertebrates like birds, fish, amphibians, and rodents, had been found to evolve and survive challenging conditions. Extinction was not inevitable. Some bat populations were surviving, Auteri knew, but was it evolution or did some simply survive in some other, nonheritable way? One way to approach that question is to compare the genetic makeup of bats that survived and those that did not. The goal would be to detect a genetic shift. Unlike the focus in the Grants' work, Auteri's was on the genes rather than the known physical traits encoded by those genes. Because she collected DNA samples from bats that had been found dead from the disease, and the bats had already started decaying, she couldn't accurately measure things like weight. But clues could still be found from their DNA. She used a sort of a genetic shotgun approach—randomly breaking DNA into fragments and sequencing them. She hoped to find that some fragments would be more frequently associated with the survivors, which would mean bats were adapting to the disease. The genes these fragments were part of could also yield their own clues and be used to look up what previous research might suggest the role of that gene was.

Auteri's first goal was to catch adult bats that were out and about, bats that she could be certain survived at least one hibernation in the presence of Pd. Some were caught in mist nets, a common practice for catching and releasing birds. The nets are strung across a field or woodland and are difficult to see even for expert navigators like bats. The fairly large mesh entangles anything that flies into it, so the nets must be checked every eight to ten minutes. For the bats nets are opened around dusk until several hours after sunset, the most active period. Other samples from active bats came from animals caught by the state's Department of Natural Resources—those flitting about in attics and homes. While the mist-net bats would be released after sampling, bats caught in homes would be euthanized and checked for the rabies virus to determine possible exposure to people or pets.

In living bats, membrane tissue is collected using a small-hole punch, the kind that might be used for a skin biopsy. Once DNA is extracted from the membrane sample, it is sequenced. Owing to the difficulties of catching and sampling live bats, only nine surviving bats were included in Auteri's study. Such a small sample size raises red flags for scientific studies and can make it difficult to detect changes, but in this case the genetic differences were remarkable. Recall that each mammalian cell has two sets of chromosomes, one set from each parent. Strung along those chromosomes are segments of DNA that comprise different genes. All together we have twenty-three chromosome pairs, or forty-six total chromosomes. Different species of bats have different numbers of chromosome pairs ranging from seven to thirty-one, or between fourteen and sixty-two chromosomes; little brown bats have twenty-two pairs. Like us, bats inherit one chromosome from each parent, which means there are two sets of DNA and that each individual has two copies of each gene. Gene variants are called alleles. We are most familiar with the different alleles for eye color, height, and skin color. Some have more variants than others. Auteri and Knowles focused on genes that had small differences called single nucleotide polymorphisms, or SNPs (pronounced "snips"). Many SNP variants are located in genes that have yet to be characterized: no one knows what the gene does, but they can still be used to detect differences. But a few SNPs used by Auteri belonged to known genes that code for known traits. Of those, four variants were associated with surviving bats and not in the others.

One of the SNPs that Auteri used was part of, or near to, genes that coded for a neurotransmitter receptor. Another was linked to echolocation, and a third to the immune response. All are responses that might help a bat survive. But one interesting gene variant is called cGMP-PK1, which has been linked to obesity in mammals. Auteri notes, "None of these alleles were in the non-survivor group. None at all. I was retrospectively surprised, if that makes sense." That any one of these gene variants might be shuffling around in a population makes sense since variation in all these qualities could be useful depending on conditions—just like genes for dry-weather beaks in the Grants' finches. Drought was a relatively rare event, but when it happened, had there not been enough genetic variation, the birds would have gone extinct. That said, understanding if and how much these genes are contributing to survival is unclear.

There are hopeful signs that other bat populations may survive as well. In 2019 scientists working with little browns in New York found that the mortality of those bats had dropped dramatically to around 50 percent from a high of nearly 90 percent. Surviving bats seemed to have prolonged bouts of torpor, allowing them to preserve their precious supply of winter fat. The little browns, they concluded, showed that a mammalian species could evolve rapidly in response to a novel pathogen. The study matched with Auteri's, which found differences in a gene regulating arousal from hibernation.

"I don't think we are seeing *increases* yet [in local populations]. But there are fewer dead bats," says Auteri.

* * *

THERE ARE OTHER WAYS A POPULATION OF BATS—OR FROGS OR other animals—might survive a deadly disease. Sometimes, as the immune system becomes better at killing the pathogen, the pathogen may evolve to become less provocative to the immune system. Or when a virus or bacteria or other pathogen doesn't kill off the entire host population, it has a better chance of surviving in the long run. This is the prevailing idea: eventually pathogens evolve to become less proficient at killing their host (though this isn't always the case).

Disease ecologist Jamie Voyles, like Karen Lips, witnessed the before and after effects of Bd on frogs in El Copê, Panama. In the early 2000s as a PhD student, she collected frogs from forest transects. Golden frogs, the iconic canary yellow and black frogs, were everywhere. "If you walked down the stream, it would be hard not to step on them," recalls Voyles, now an associate professor at the University of Nevada in Reno. "And you could see what a rainforest looked like with fifty to seventy different species a night." She had managed to get through about a year of frog collecting when Bd arrived. So she turned her attention to other things (dead frogs don't make a good dissertation) until about a decade later when she returned to the same site. She put together a team to survey the area for the golden frog. "Maybe it was the right time, right place," she says, but there were frogs. The first one took about two months to find. "We took about a zillion pictures!" And that, she says, is when she had her lightbulb moment. After such devastation why were there any survivors at all?

Voyles was pretty sure that the Bd fungus had become less of a killer.

But after comparing archived samples of fungi cultured from frogs ten years apart—at the beginning of the epidemic and then later—Voyles and colleagues found no evidence that Bd had evolved into a kinder, gentler pathogen. Instead, it seemed that the frogs had changed.

PANAMANIAN GOLDEN FROG

When she had first started her research, before the frogs disappeared, Voyles had sampled frog mucus or skin secretions and archived them. The secretions contain some of the frogs' frontline immune responders, which allowed Voyles and others an opportunity to compare the immune response of frog populations that had never seen the fungus with those of surviving frogs decades later.

In some cases frogs from surviving populations were two to five times better at limiting the growth of Bd. Rather than the pathogen evolving to become less virulent, it seems the frog hosts had evolved to become more resistant. "My guess, were I a betting woman," says Voyles, "is that there is standing genetic variation, and then an intense genetic sweep. Mother Nature is incredibly creative, and with time my hope is we'll continue to see amphibians recover."

Disease ecologist Vance Vredenburg, who witnessed the devastation at his Sequoia–Kings Canyon field site, sees hope in other regional populations. In Yosemite National Park he found Bd-infected frogs inhabiting hundreds of little ponds. It is a sign that the disease has become endemic; frog and chytrid have coevolved. One hypothesis is that the skin microbiome has changed, and there is an increase in microbes producing antifungal chemicals. But microbiomes are complicated, and changes in them can be difficult to interpret. There is likely more to it, but Vredenburg adds, "This really gives me hope, although at Sequoia–Kings Canyon they are still getting hammered, there are a few that will survive. *If* we can let evolution happen, these animals have a good chance." This may also be true for the little brown bats, some of which seem to be holding their own—if only they can hang on long enough for populations to thrive again. Like the Grants' finches, one of these species' greatest assets is genetic diversity within the population. For these animals perhaps evolution will come to the rescue given the right conditions and enough time. It is a big if, but there are glimmers of hope.

Part II

RESOLUTION

Chapter 6

RESISTANCE

It is a late summer day in the Northern Cascade mountains. The air is alive with pollen, dust, and spores, millions of spores released from molds and mushrooms and other fungi. One is a basidiospore recently released from the underside of a currant leaf by *Cronartium ribicola*, the fungus that causes pine blister. The spore travels with the wind for a few hundred meters or so before dropping from the sky. It could land on anything: the windshield of a car, a patch of heath, or even the soil. All would be dead-ends for this fungus. But this particular spore lands on the needle of a whitebark pine. It's a successful landing. Hyphae emerge and probe the green needle seeking a way in through the stomate, but the fungus does not get far. The pine leaf's cells in the infected region quickly die off. The fungus that feeds on the living and is surrounded by dead and dying cells is stymied. The spore has landed on a pine that is able to resist its advances through a type of protective cellular death. The tree's ability to fend off disease owes thanks to a set of genes that provides the pine some protection against a potential killer fungus. The genes have been conserved over thousands or millions of years through the process of natural selection just as with the genes that shaped the

Grants' finch beaks or possibly the "fat bat" genes. Maybe eons ago ancestors of the trees and fungus once met. Or maybe the genes were conserved for some reason other than a fungus, to protect against some other invader or environmental condition. Scientists believe that the rust-resistant tree and others like it are the pine species' best hope against this fungus.

Once a fungal pandemic takes hold in the wild, whether in bats, frogs, or trees, most scientists agree that there is no "going away." The once-novel fungus remains in the forest or pond or soil, infecting generation after generation of host and in some cases remaining long after the host is gone. This means that if a tree or a bat or a frog is to survive, it must find a way to live with a potentially deadly fungus, which in turn depends on at least two things: the existence of useful genes in the surviving population and a means of dispersing those genes into the broader population in time to save it. For trees, which can take decades to mature, if we let nature take its course and wait for resistance genes to spread throughout a population challenged with a fast-moving killer fungus, we may risk the loss of those infected populations if not an entire species.

When commercial shipments of pine seedlings unleashed blister rust fungus a century ago, the fungus killed millions of five-needle pines across the United States. Some surviving trees, including white-bark pines, western whites, and sugar pines, were lucky. Maybe they lived in forest pockets where the fungus hadn't yet penetrated. But survivors living among dead and dying trees suggested some level of natural genetic resistance to the non-native disease. For centuries plant breeders have long appreciated the power of a useful genetic trait (long before anyone understood the underlying genetics), selecting, breeding, and growing plants that survive disease or drought or grow sweeter fruit or fewer seeds: evolution aided by the human hand. During the days of *Ribes* eradication, when young men scoured the wood for currants and gooseberry shrubs amid areas hard hit by the fungus, they noted here and there the odd survivor, a sugar pine (*Pinus lambertiana*) or a western white pine *(Pinus monticola)* standing tall. In the 1950s, when it was clear that controlling white pine blister rust at least in the West was futile (*Ribes* control continued into the 1970s in Yellowstone National Park), a handful of foresters wondered if those surviving trees were genetically resistant, and if they were, was it possible to breed them like any other plant? The *selection* and *breeding* of rust-resistant trees would be something new. The life cycle of agricultural plants can be counted in seasons or a year, but the life cycle of trees was decades or more. So while breeding a new crop plant might take years, breeding disease-resistant trees could take decades. The first candidates for blister rust resistance breeding programs were western white and sugar pines, trees not only valuable for timber but admired for their majestic appearance. For the scientists devoted to saving them, it would be a moonshot, and any successes might not be realized for decades. But it was a challenge worth taking.

⬤ ⬤ ⬤

WESTERN WHITES LIKE THEIR EASTERN COUSIN GROW TALL AND straight and make good wood. "The tree presenting then a more slender and elegant shaft than any other tree in the woods," wrote the naturalist John Muir of the pine, which could tower 150 to 200 feet above the forest floor. Like the eastern pines, the western white pine dominated forests from Oregon to California, covering two million acres in Idaho alone. Its light, clean, easily milled wood was a resource too good to ignore. By the turn of the last century sawmills quickly consumed entire forests, turning the so-called king pine into billions of board feet (a "board foot" is a unit of lumber measurement based on a twelve-inch-wide slab of lumber). Sugar pines are considered the largest pines in the world, and two of the tallest and broadest of this species live around Lake Tahoe, California. They are ten yards short of a football field in height, and their trunks are broad enough around for five or six adults to join hands. The sugar pines were an obvious target for mid- to late nineteenth-century settlers whose mills made commodities of the trees, setting off what one writer has referred to as a "whirlwind of destruction." Sugar pine wood built gold rush homes and propped up mines while providing structure to sluices and flumes through which gold-dusted water ran. At one point early in the twentieth century the US Forest Service estimated some thirty-nine billion board feet of sugar pine stood in California's forests. What foresters and loggers could not have imagined was that blister rust spores were hopscotching their way from British Columbia down through the Cascades. The favored trees were in trouble.

In 1946 Richard Bingham, a forest pathologist, was a junior member of a blister rust control program operating out of Spokane, Washington. His job included surveying rust-damaged pine forests. That year he came across a remarkable find: a sixty-year-old, one-hundred-foot-tall (thirty meters), healthy western white pine. Over the next few years he spotted fourteen more healthy trees surrounded by dozens of rust-killed hulks. The trees, he surmised, were able to resist or tolerate the rust infection. Bingham knew that scientists in the eastern states who were experimenting with rust-resistant eastern white pines had shown that resistance appeared to be under genetic control. In other words, a resistant parent ought to be able to pass whatever genes provided such tolerance on to its offspring—*if* the trees could be successfully bred. Bingham would go on to lead a program devoted to breeding blister rust–resistant western white pines. With no illusions about the time involved and the difficulties ahead, the breeding effort was slated for a five-decade run.

The blister rust resistance program would stretch from the Northern Rockies to the Pacific Southwest. Other scientists, geneticists, foresters, and breeders joined and expanded its efforts to other blister rust–susceptible species like sugar pine. No one knew exactly *how* the trees managed to survive or for how long they could fend off disease. But they knew that many seedlings grown from resistant trees appeared to have inherited resistance at least for a period of time. A tree might resist at first but die ten or twenty years later, or the fungus might evolve to counter a tree's particular resistance, negating decades of work. The process was both scientifically and physically challenging—and in the early days life-threatening. Western white and sugar pines set their cones near their crowns, sometimes well over one hundred feet aboveground. Breeding the trees required foresters to hand-pollinate seed-bearing cones and, when the seeds had matured, collect ripe cones. Gerald Barnes, the teen who had tramped backcountry pulling *Ribes,* joined the US Forest Service's tree-breeding program in 1962. One of his first tasks, he recalls, was to climb a sugar pine. He had accompanied the program's regional geneticist, Tom Greathouse, on a trip to pollinate the trees. They stopped before one particularly towering tree, and Barnes watched through a pair of binoculars as the geneticist climbed the first thirty feet to the lowest green limbs. He used a contraption called Swiss tree grippers, or a tree bicycle. The grippers allow climbers to make their way up limbless trunks without causing too much damage to the bark, which matters when the same tree must be scaled repeatedly. When he reached the limbs, Barnes watched as Greathouse free-climbed dozens of feet more—no grippers, no ropes—into the green. When he came down, it was Barnes's turn. His task was to retrieve the red ribbon Greathouse had tied to the very top of the tree. A nervous Barnes recovered the ribbon. Over the next twenty years he climbed some of the tallest pines in the West, pollinating and retrieving pine seed cones and checking trees for rust cankers. Once he was in a treetop during an earthquake. Another time, a delta-winged aircraft swooped overhead close enough to rock the tree. He climbed through cold and snow and nearly died when, approaching the crown of a 165-footer, he jumped onto a branch that broke. He made it down with only a broken collarbone. Barnes and Greathouse subsequently wrote the first informal guideline for tree-climbing. The risky work of climbing trees has changed some over the years, but cone collection remains as important today for tree-breeding as it was in the program's earliest days.

• • •

FOR MUCH OF HISTORY MOST TREES HAVE REMAINED RELATIVELY wild and true to their genetic roots. Today's sugar pine or western white would still be recognizable by those who lived with the trees centuries ago. The same cannot be said of food crops. For millennia farmers shaped what we eat from wheat to bananas, potatoes to cabbage, even if they didn't fully understand the inner workings of heredity. Actively harnessing the power of genetics and selecting for genes for better taste, faster growth, and drought and disease tolerance is a relatively recent tool. A better connection between plant breeding and inheritance likely originated in the mid-1800s with biologist Gregor Mendel, an Augustinian monk. Mendel bred peas with characteristics like green or yellow color, wrinkled or smooth seeds, and he noticed that traits were transferred from one generation to the next in generally predictable patterns. The smooth seed trait, he found, was dominant—as were yellow seeds and green pods. For traits controlled by a single gene (pair) like those studied by Mendel, one allele or gene variant is donated by the father and the other gene by the mother. A dominant allele of a gene masks the expression of the other allele. For a recessive trait to emerge, two recessive alleles, one from each parent, are required (although *clear* cases of dominance are the exception rather than the rule). Mendel's work was revolutionary and would someday provide breeders with the knowledge to better direct the outcome of their efforts, but in his lifetime few scientists paid any attention. Most of us are introduced to Mendel's laws of genetic inheritance in school. We look at the eye color of our parents and understand why two blue-eyed parents have all blue-eyed kids and why the kids of two brown-eyed parents can have brown, blue, or hazel eyes. Even as genetics becomes increasingly complicated, for single gene inheritance, Mendel's work has stood the test of time and provides us with a fundamental understanding of inheritance and gene dominance. The laws of inheritance from Mendel's work provided plant breeders with a powerful set of tools.

As the twentieth century progressed, scientists discovered that traits are encoded by genes carried on chromosomes made up of genetic material called DNA. All of these findings provided deeper insights into heredity and how genetic material is parceled out one generation to the next. Scientists have also found that while some traits are controlled by different versions of a single gene, most traits are controlled by many genes, which makes understanding the inheritance of certain traits complicated. It was fortunate for Mendel that he worked on pea plants which had simple single gene effects. Had he picked another garden plant, inheritance may have been harder to figure out. At least fifty genes and different regions of the human genome are 80 percent responsible for determining height. Back in the 1950s, when studies were under way to understand blister rust, no one knew how many genes contributed to resistance. Was it a single dominant gene or many genes acting in concert? Which was better? Evidence from food-crop breeding showed that disease resistance conferred by a single or a few genes is easier to capture and highly protective but can be ephemeral. If there are only a few hurdles, a fungus or another pathogen might quickly evolve around resistance, negating years or decades of effort. Breeding for multigene resistances is far more complicated and might only confer partial resistance, although it is likely longer lasting.

One of the first big breakthroughs in pine breeding for resistance came in 1970 when Bohun Kinloch, a US Forest Service geneticist, and his colleagues reported that they had found a single major gene that protected sugar pines. The results were textbook Mendelian genetics and followed the patterns of dominant and recessive genes. While foresters were uncertain how long this kind of single gene immunity would last, they at least had a gene to latch onto. Kinloch would later find multiple gene resistances as well. The gene was dubbed *Cr1*, and when resistant needles are infected, the plant tissue around the new infection dies—a cell suicide response that prevents rust from spreading. Another single resistance gene, *Cr2*, was found in western white pine. In both cases, though, strains of rust had long since evolved that overcame those single gene resistances.

The breeding programs that started in large part with Richard Bingham have been ongoing now for more than seventy years. The goal for breeders today is to find trees protected by both major genes *and* multiple genes, a complementary protection and more difficult hurdle for blister rust to jump.

Throughout the 1900s blister rust wasn't just attacking the western white and the sugar pine; it killed the whitebark pines too. But that tree, of little value to the timber industry, hadn't been included in the early restoration efforts. Leaving this ecologically important species' future to nature was a gamble ecologist Diana Tomback and her colleague Robert Keane, who at the time worked for the US Forest Service, weren't willing to take. In the mid-1980s, fearing the loss of the whitebark, they and others including Kinloch were part of a research team of different experts. Their aim was to determine all the reasons whitebark pine populations were declining. They found that blister rust was one of the major causes, and by the late 1990s there was substantial agreement that whitebark needed the same chance that had been afforded their five-needle cousins like the western white and the sugar pine. This coalition of passionate scientists, managers, and other professionals would urge the development of new resistance breeding programs. Richard Sniezko, a geneticist who works out of the USDA's Dorena Genetic Resource Center, independently joined the "resistance" effort. Their work would move the effort beyond saving trees primarily for commercial value to saving trees for their intrinsic value as well.

Surrounded by the Willamette National Forest to the east and the Umpqua National Forest to the south, the Dorena Genetic Resource Center is situated in a paradise of pine and spruce set amid snow-capped mountains and clear lakes. Sniezko's passion is saving trees by developing resistant populations through breeding. Much of his work centers on identifying potential rust-resistant trees, confirming resistance, and then ensuring enough genetic diversity is maintained in restored populations. When he first arrived at Dorena in the early 1990s, the center's rust research program focused on restoring sugar and western white pines. Restoration has always required a good deal of strategy. It would be relatively easy, he says, to find resistant trees (referred to as "mother trees"), collect buckets of their seed, and plant the seeds. But there might not be enough genetic diversity over the long haul for these new populations to survive future disease or environmental challenges, or seedlings grown from a rust-resistant mother tree located in the north may not fare well in the south. A tree growing at sea level may not thrive at altitude. The same is true for those that have adapted to a wetter environment compared with a dry climate. When climate change is considered, trees from the south may be needed for a warmer north. So the trees from which seed is collected must be representative of each region, north to south, and from high elevation and low.

The center now evaluates mother trees of several white pine species, including whitebarks, for genetic resistance to blister rust. Like other pines, whitebarks set cones high above the ground. This requires multiple visits during the season by skilled tree climbers (although unlike the western white and sugar pines, whitebarks are shorter in stature and so are more manageable). Climbers scale the trees to cage candidate cones in June, protecting them from animals seeking fat-rich pine seed. In late summer the cones are assessed and collected. Whitebarks are wind-pollinated, meaning their seed cones are fertilized when the smaller pollen cones—resembling a cluster of red or purple berries—ripen. If there are pine trees near your home, you've likely witnessed the enormous amount of pollen released on a windy day. Much of it clings to our windshields, infiltrates our homes, and floats across field and forest; some of it lands on seed cones. Cones from a single mother tree will likely be fertilized by pollen from several other trees.

In their first year seed cones, like pollen cones, appear small and red, but they sit higher up in the tree's crown—a common strategy in trees to reduce the chance of self-fertilization. Once fertilized, the cones and their seed require fourteen months before they are fully mature. Collectors seek seeds with healthy pine embryos, which they reveal by slicing a few sample cones from stem to tip. Sometimes a cone looks healthy, but the seeds have been eaten by insects or the seeds haven't been pollinated and there isn't a healthy embryo. The cones are processed at Dorena, where seeds are removed and prepared for storage or planting. Because the cones evolved to be pecked apart by nutcrackers, removing the seeds from the tightly closed cones can be a challenge. Dried cones are tumbled together with rubber balls to smash them apart without damaging the seeds. They are then processed to separate seed from bits of smashed-up cone. Digital X-rays help separate the good seed—those with healthy whitebark embryos inside—from the damaged or underdeveloped. Several dozen seeds are typically recovered from a good cone. Most are stored in freezers in carefully labeled packets so they (and the mother tree from which they came) can easily be tracked down when needed. Waking the seeds from their dormancy and readying them for resistance testing is a separate process that takes months of soaking, warming, cooling—essentially re-creating what a pine seed nabbed by a nutcracker and cached in soil at altitude might encounter. Storing and preparing seed is both time intensive and labor intensive. And that is just the first step in resistance breeding.

By 2021 the Dorena Center had tested seedlings from approximately 1,500 whitebark mother trees from Oregon and Washington State. Because there isn't yet any kind of rapid test for resistance genes, identifying rust-resistant trees follows the same protocols developed fifty years ago to test for resistance in white and sugar pine. The whole process from seed collection to resistance detection can take up to seven years. In preparation for each trial, over 60 seedlings from 120 promising mother trees are grown out until they are two years old—short, green toddlers bristling with needles. When it is time for testing, some 7,000 seedlings are loaded into a large, garage-like "fog" chamber and showered with fungal spores. Because whitebark needles in nature are infected with spores released from infected *Ribes* leaves, Sniezko's team uses naturally infected leaves as a source of spores. Staff members collect thousands of leaves speckled with orange rust and load them onto mesh racks suspended just above the pines. Because the spores require moisture to germinate, fog is pumped into the chamber. The spore-laden fog chamber with its bristling young trees is an idealized version of nature that can be controlled with more, or less, fog or spores. Once the fog clears, the saplings are moved outside and into raised beds where the infected trees are allowed to grow.

Over the course of the next five years most will die from their infection—some quickly, others over time. Some may get sick but recover. A few will survive. Each mother tree is scored according to the performance of the offspring. Seed from the most-resistant parent trees are granted an A or B rating based on the resistance of their progeny. These trees are deemed suitable for restoration, although in high rust areas, sometimes only 50 percent of an A-rated tree's progeny are expected to survive. The rest are graded C through F, depending. Each year perhaps only about 10 percent of the parent trees rate an "A."

After years of testing Sniezko and his team now have a better idea of the patterns of resistance throughout parts of the Northwest. There is map of whitebark habitat from Washington State and Oregon he likes to use to illustrate the distribution of resistance. It is overlain with small, colorful pie charts. The colors represent degree of resistance: green, blue, yellow, orange, and red. Green represents mother trees with an A rating; failures are red. There is plenty of red scattered around the Northwest map, particularly in eastern Oregon, but there is green around Mount Rainier National Park in west-central Washington State.

Crater Lake National Park in southwestern Oregon shows up on the map as half green and blue. In 2006 the park relocated a parking lot and created a newly opened space. Sniezko and park personnel saw the space as an opportunity to plant resistant seedlings. When trees are planted in the backcountry, they are monitored but can be remote and difficult to access; these new trees would be out in the open and easy to check on. Their highly visible location could also provide the public with some conservation-restoration education. Sniezko's team grew the seedlings, and park personnel planted some three hundred little whitebarks, just a few inches tall, each protected by small piles of rocks. Forty seedlings were the offspring of an A-rated family; the rest of the seedlings were a mixture representing C-rated trees. Sniezko says the trial used seedlings available at the time. More recent trials use offspring primarily from A- and B-rated families. By 2021 the tallest trees in the old Crater Lake parking lot stood about six feet. Some have rust, and at least one died. But most, Sniezko says, are looking "really, really good." It will be decades before anyone knows for sure how resistant they are—not exactly reforesting but a place to begin and a bit of hope for the whitebark.

A healthy forest is a dynamic landscape where mixtures of trees and shrubs jostle for dominance over long expanses of time. A species may rule the overstory for decades until a storm knocks them over or snaps a trunk, leaving behind a crownless snag. These sorts of changes allow less shade-tolerant trees, which have bided their time in the understory, to take their place in the overstory. Some release chemicals encouraging certain neighbors and discouraging others. Increasingly, for a number of reasons ranging from the changing climate to forest management practices, wildfires sweep through, leveling massive numbers of trees, which makes reforestation complicated even in the best of circumstances. Whitebark restoration across its range requires more than simply planting rust-resistant seeds or seedlings. It needs enough nutcrackers to cache and distribute seeds, along with consideration of how the plant community has changed, what other conifers and understory shrubs have grown in since the whitebarks died, and whether there is a risk of fire. Whitebarks are some of the earlier colonizers after a burn because they have been planted by nutcrackers. But the trees are also highly susceptible to fire.

Restoration, along with conservation and reforestation, is an art and a science, particularly if the intention is that the trees conserved or planted today will still be standing a century or two from now. Bob Keane has considered this for much of his career, particularly the relationship between whitebark pines and fire. Fire is both a cause and consequence of changing ecosystems and sometimes a tool for those replanting trees. Prescribed fire and other techniques may ensure that shade-shy whitebarks will thrive instead of getting choked and outcompeted by other trees, though fire can also kill rust-resistant trees, particularly young trees. A newly planted mountainside swept by fire will most likely lose its young pine trees. Climate change has increased fire risk. Soil type and soil microbes are important too. Like most other trees, whitebark pines are dependent on soil fungi, which help the trees gather nutrients, stave off drought, and tame other soil microbes. The understanding of the role of certain soil fungi as helpful if not essential members of the tree's soil microbiome is gaining momentum. There are now soil inoculums specific for newly planted whitebark pines, and some scientists suggest that seedlings ought to be planted near surviving trees or where whitebarks once lived so they might benefit from the existing soil microbes.

Hundreds of thousands of whitebark pines have now been planted throughout their natural range by those invested in the trees' future, encouraged by organizations like the Whitebark Pine Ecosystem Foundation and the American Forests conservation nonprofit. Some whitebarks, planted in the days before the resistance breeding programs, have died, but now seed and seedlings from resistant trees are used. There is the potential for hundreds of millions of trees to be planted across nearly two million acres of habitat. The aim is to plant seedlings from resistant trees, but that will depend on seed crops, cone collection, and the whitebarks themselves. In 2012 Keane, Tomback, and others drafted a multiagency whitebark pine restoration strategy, and Tomback, representing the Whitebark Pine Ecosystem Foundation, in partnership with American Forests and others, are working on a multiagency National Whitebark Restoration Plan. Its foundation rests on the decades of research by scientists including Tomback, Keane, and Sniezko in fields spanning forest ecology, geography, and genetics. Select live, healthy populations would be identified and protected, while genetically resistant trees will provide stock for replanting. Both the existing and the planted trees are critical to the species' survival. This begs the question, in the age of 23andMe and rapid genetic testing, is there a more efficient route than the seven to twenty years now required to identify and vet resistant trees?

• • •

IN 2016 A LARGE CONSORTIUM OF SCIENTISTS INCLUDING DAVID Neale, a plant scientist and geneticist at the University of California, Davis, sequenced the sugar pine genome. A genome is the collection of genes that encodes living things: a human, a mouse, or a tree. A whole genome, depending on the species, can be like a huge, wordy book. And the collective expression of traits, the translation of this

book into a mouse or a tree or even a pea plant, is far more complicated than Gregor Mendel could ever have imagined. For much of the
twentieth century scientists could only read a handful of words in the
genomic tome. They could sequence and identify a specific gene, but
without the surrounding genes and genetic controls there was no con-
text. In the 1970s breakthrough discoveries by Nobel laureates Walter
Gilbert, Fred Sanger, and others enabled scientists to sequence faster
and read more. Still, they were lacking the whole story, and even
within the paragraphs some words remained garbled or missing or
out of order. Before the computing power available today, the work
was tedious and expensive. But by the close of the twentieth century,
not only could scientists read whole sentences and paragraphs but—in
some cases—entire books.

In 1995 scientists decoded for the first time the entire genome
of the bacterium causing a type of pneumonia; then they sequenced
the genes of a nematode. In the early 2000s, when the human
genome with its twenty thousand to twenty-five thousand genes
was finally sequenced, scientists were surprised that such a small
number of genes makes us who we are. By length and weight our
code boils down to around 205 centimeters and just about 6.4 pico-
grams (one picogram is a trillionth of a gram) depending on our
sex. Female genetic material is slightly longer and heavier. Full
sequencing of pathogens, laboratory animals, model plants, popular crop plants, insects, and fish followed. It was only a matter
of time before the genomes of trees would be sequenced. Black
cottonwood was the first, followed by flowering trees with small
genomes like peach; Norway spruce, white spruce, loblolly pine,
and Douglas fir followed. Sugar pine was the first white pine to be
sequenced. With thirty-one billion base-pairs (subunits of DNA
that form each "rung" of the DNA ladder), the genome was the
largest ever sequenced and assembled. For comparison, the human
genome consists of about three billion base-pairs.

Finding and linking a trait like resistance, which may result from the expression of multiple genes working together, is far more complicated than identifying a single dominant gene. One way to accomplish this is to use genome-wide association studies, or GWAS. When the human genome was sequenced, geneticists used GWAS to link genetic variations with some diseases. It is a data-heavy process that compares whole genomes against one another for differences one gene locus at a time. Sequencing the genome of an organism, says Neale, is like making a parts list. The next step is to sequence a whole bunch more and find variation in the population. The third step is to link genetic variation with the expression of traits or characteristics of interest. Imagine, he says, how this might be done in humans. Take one thousand volunteers, screen them for disease susceptibility, and rate that susceptibility on a scale of one to ten. Each individual is then genotyped for all twenty-five thousand genes, and their entire genome, gene by gene, is compared with their susceptibility score. Then each individual is compared with all the others. Most comparisons will have no connection, meaning the gene had no or little influence on disease. Every so often a gene in those individuals correlates with disease susceptibility. The results don't come cheaply in dollars or effort. Early on, just sequencing the human genome cost between hundreds of millions and a billion dollars and took years. Sequencing the genome of an individual now costs under one thousand dollars and can be done in days.

David Neale's next target is the whitebark pine. In 2020 the US Fish and Wildlife Service proposed adding the whitebark pine to the nation's threatened and endangered listing. If granted, the listing could help move the genetic sequencing along. Working out the genome and identifying specific genes for any given tree is a big project. Still, the day may come when a forester can drop a pine needle into a tube and within hours know if and how the tree resists rust.

The genomic approach is a powerful technology. It could help expedite the current testing efforts. Once whitebark pine resistance genes are identified, more rapid DNA testing could point researchers to the most promising mother trees, whose seed would then be put through the rigors of the rust challenge. The combined approach would knock years if not decades from the traditional approach.

The lessons learned from the whitebark pines will help other forest trees affected by non-native pests and pathogens, too. Modern genetics (genomics) will speed up the selection process, but trees grow, mature, and reproduce across timescales vastly different from ours. Because restoration and conservation efforts will play out over decades and centuries, few people alive today will be around to enjoy whatever successes may come. Even if scientists restored 20 to 30 percent of the trees with genetic resistance, the ultimate success, according to Tomback, will rest on the nutcrackers. Hopefully, they will not turn away from the once reliable food source. Success will also require a new generation of scientists, conservationists, and citizens as devoted as Tomback and her colleagues.

Trees are some of the longest-lived and oldest organisms on Earth. Some live thousands of years. As species they survive natural disasters, climate variation, the cycling of insects, and forest fires, and adaptation to these challenges are written into their genomes. Some genes might encourage a tree's seed to grow well in burned soils after a fire. Others might provide defenses against an outbreak of beetles. But this natural process of selection may not lead to adaptation and survival if there is a sudden and novel challenge. And yet this is the situation into which we have put these trees—the chestnuts, the whitebarks, and countless others, including elm trees, eucalypts in Australia, and ʻōhiʻa in Hawaii. As with the five-needle pines, when there is sufficient existing genetic diversity, trees can be saved through the process of unnatural selection—the intentional breeding by the human hand. This is a method farmers have been using for centuries, and it is one of our greatest tools.

Chapter 7

DIVERSITY

In 2003 agriculturalist and sociologist Cary Fowler and his colleague Henry Shands had an idea. They would save the world's plant diversity. Both had worked in agriculture and were familiar with the global food culture and economy. They knew better than most the critical role of good seed that readily germinates into healthy plants—even after years in storage—in feeding the world. And they knew the consequences of the loss of genetic diversity in crop plants. A seed, which protects and provides for a plant's embryo, holds the necessary DNA to propagate a species. It is the most basic unit of virtually all the food we eat (except bananas and other seedless wonders).

For much of our agricultural history, breeding and improving food plants had been an inexact process. Deep down in a cultivar's genome there had always been some wildness, providing enough variation in a field of wheat or rice that should some plants die from heat or salt or insects, others would survive. These domesticated crops that retain genetic diversity are called landraces. These were the dominant crops throughout much of our agricultural history. In the twentieth century plant breeders, armed with a better understanding of inheritance, became increasingly effective at breeding plants to their liking, keeping desirable traits while breeding out others and effectively putting crops through the genetic wringer. Now many landraces have been replaced by highly cultivated plants that are more genetically uniform and predictable. As growers moved away from landraces, the plant's gene pool narrowed. We got better-tasting, prettier, more abundant fruits and vegetables, but the cost was genetic diversity.

Harvests increased by the ton, while genes dropped by the wayside. Popular crops narrowed from hundreds or thousands of species to the handful of best performers. In the United States the wheat, corn, and tomatoes that grew fast and resisted pests and pathogens could more easily be shipped across the country or even farther. It was all good, except that between 1903 and 1983 over five hundred varieties of cabbage narrowed to a little over two dozen, we lost almost all of our peanut diversity, and we abandoned more than three hundred varieties of tomatoes. In many fruits and vegetables beneficial characteristics for the plants, such as protective chemicals that might discourage pests and pathogens and predators—including human predators—were bred out in exchange for better-tasting tubers, roots, leaves, and fruits. As crops became more and more uniform, the risk of massive crop failure from disease, pests, or climate increased. This didn't happen only in the United States but around the world.

Over the past century plant breeders, scientists, hobbyists, and farmers began to worry about the loss of favorite crop varieties and the loss of landraces, so they began collecting and preserving seed and other genetic material. Some of them save seeds because they enjoy a good heritage breed. Others do it because they believe that if they do not, one day so much diversity will be lost that there won't be enough food for the world to eat. In the early 1990s the Food and Agriculture Organization of the United Nations (FAO) enlisted Cary Fowler to assess the current state of the world's crop diversity. Fowler had worked in agriculture and was familiar with the global food culture and economy. "What my team and I found," wrote Fowler, "was shocking." The world's most valuable natural resource—the germplasm for the food we eat—was in peril. Several years later he and Shands would propose developing a backup system for the world's plant diversity, ensuring viable good seed for future generations.

In 2006 the Global Seed Vault in Svalbard, Norway, began collecting its first seeds; Fowler spearheaded the project. Some call it the Doomsday Vault, although Fowler doesn't like the name. The vault sits about seven hundred miles from the North Pole. The entrance is reminiscent of some science-fiction portal, a futuristic doorway set into a foreboding landscape that opens into a refuge from what has become an unlivable planet. Only in this case it is a refuge for seeds— hundreds of millions, potentially billions of seeds. The doorway leads to a 130-meter-long horizontal tunnel carved into rock that opens onto a space coated in glistening ice crystals. Fowler calls this the cathedral. There, ice-encrusted doors lead into three separate rooms protected by heavy metal doors where temperatures hover around −18°C (−0.4°F). It is a giant, bombproof freezer.

In the early twentieth century the largest collection of seeds in the world was housed at the All-Union Institute of Plant Industry in what was then Leningrad (now St. Petersburg), Russia. Most of those seeds were collected by Nikolai Vavilov. Like the USDA's agricultural explorers David Fairchild and Frank Meyer, Vavilov traveled the world seeking edible crops. He also studied genetics and understood the value of plant genes for disease immunity. Russia had plenty of experience with starvation, and Vavilov intended to prevent future famines by creating and curating the seed bank. By 1941, when Adolf Hitler laid siege to Leningrad, starving its two million citizens, the bank held seed from over three hundred thousand plant varieties. There were potatoes, rice, corn, and wheat, all plant matter that was, incidentally, edible as stored. Starving locals knew this, as did most likely the Germans. Several Russian botanists who were determined to protect the invaluable cache locked themselves inside surrounded by food they would not consume. The siege lasted until 1944. Between 1942 and 1943, while preserving seeds for future generations, at least nine botanists died from starvation—having refused to raid their valuable stores. Vavilov died in 1943 while serving out a twenty-year term in a work camp after having been arrested and accused of being a British spy. His cause of death was also starvation.

Beginning largely in the 1960s and 1970s, seed banks began to emerge around the world. By 2010 nearly two thousand large and small banks held over seven million seed samples. Some specialize in a few crops or are able to store only a limited number of seed; others are immense. There is a collection of rice at the International Rice Research Institute in the Philippines. The International Center for Wheat and Maize is in Mexico, and hundreds of other such collections are scattered around the world. In the United States the campus of Colorado State University in Fort Collins houses a seed vault known as the USDA Agricultural Research Service National Laboratory for Genetic Resources Preservation. The vault is part of the USDA's National Plant Germplasm System, which includes a site in Pullman, Washington, that stores germplasm for alfalfa, chickpeas, and lettuces. Maize, millet, and quinoa are stored in Ames, Iowa. A station in Geneva, New York, conserves the germplasm of apples, cherries, and grapes. The Fort Collins site is in part the backup location where seed and germplasm from other sites are sent for deep storage in a catastrophe-proof concrete building. In the 1990s the facility began holding the nation's animal gene bank, and in the 2000s it added viruses, fungi, and bacteria for research. The USDA bank now holds germplasm from nearly 13,000 plant species, mostly as seed but also as roots, shoots, and dormant buds. There is also semen, blood, and other bits of DNA from cows, salmon, honeybees, and screwworms. There are seeds and germplasm from endangered plant and seed conservancies, too, including seed from whitebark pine. This is, in effect, the nation's food vault. It, along with the others around the world, holds what one USDA scientist calls "the thin green line" between food security and global starvation.

And yet, back in 2003, Fowler and Shands feared that few of the banks were truly safe. Many were underfunded and unable to purchase what they needed; others were situated in politically unstable regions with little or no security, or their freezers weren't reliable. "Many genebanks were not so much banks," observed Fowler, "as they were hospices. A few were morgues." And so they devised their plan B. By 2015, millions of seeds, from nine hundred thousand unique samples submitted by over two hundred different countries, were in storage at the Global Seed Vault. As the acquisitions climb each year there is plenty of room. The vault is designed to hold two and a half billion seeds from over four million crop varieties. If there were to be some global cataclysm, if crops were to begin dying from the rapidly changing climate or, say, a widespread fungal infection, and war or other environmental disaster wipes out existing seed banks, the future of humanity might well be stored in the cold, deep inside of a mountain.

* ● *

THROUGHOUT MUCH OF RECORDED HISTORY ONE OF THE MOST feared crop diseases has been wheat stem rust, *Puccinia graminis*. The fungus has haunted farmers for millennia. Because some stages of the fungal infection appear deep red, the Romans sacrificed red foxes, dogs, and other animals hoping to appease Robigus, the god of rust. Today, wherever wheat grows, stem rust remains a risk. Rust caused major epidemics on farms in the United States in 1916 and then again in 1935. Like other rusts, including white pine blister rust, stem rust has another host, the barberry bush. And so, following the 1916 outbreak until 1970, barberry bushes were eradicated by the hundreds of millions across the most susceptible states. In some regions of the world that were less food secure, rust epidemics caused major famines. Rust was a danger to humanity. So in 1944 the Rockefeller Foundation sent Norman Borlaug, a young plant pathologist, to Mexico, where as in other wheat-growing regions rust was rampant. Borlaug's job was to improve the wheat grown there and to train Mexico's next generation of farmers.

When Borlaug was a college student in 1933, he witnessed a riot over milk. Milk prices had fallen, and in some places milk workers went on strike—tipping over milk trucks and beating anyone who got in their way. Nearby, hungry citizens just wanted food, and when they surrounded a milk truck on the streets of Minneapolis, a brawl erupted. Seeing the desperation of hunger, he says, eventually set him on a course that would change agricultural history. After graduating from college Borlaug fell in with a plant pathologist who studied stem rust. When Borlaug arrived in Mexico, he realized that stem rust was such a constant that growing wheat was essentially "an exercise in rust management." Borlaug rolled up his sleeves and set to work breeding wheat to be more resilient.

After years of crossing different varieties Borlaug captured a handful of favorable genetic characteristics. One important trait was resistance to stem rust, and he was able to succeed because there were enough wheat varieties to breed one into another and because some of those plants held the genes for rust resistance. Borlaug's wheat fed billions, and for that, in 1970 he was awarded with the Nobel Peace Prize. One of the genes Borlaug's wheat relied on was called $Sr31$. Now decades later, 700 million tons of wheat carrying various resistance genes including $Sr31$ are grown on 220 million hectares around the world. (The new wheat varieties were more productive than earlier strains but also more reliant on fertilizers and pesticides.)

Genetic diversity saved the crop, but the wheat gene pool has narrowed. Over 90 percent of wheat grown in the world is some variety of common bread wheat (*Triticum aestivum*), and most of the remaining varieties are durum wheat (*T. turgidum* ssp. *durum*). Many of those crops rely on the protection of the *Sr31* gene.

In 1998 a highly virulent strain of wheat stem rust emerged in East Africa; it was dubbed Ug99. The fungus was able to overcome resistance conferred by *Sr31*, and scientists and growers worried that the fungus would cause a global pandemic. When Ug99 first emerged, Borlaug wondered if past successes might have contributed to the looming disaster. A rust epidemic was like a wildfire, Borlaug wrote, and all an epidemic needed was fuel, the widespread "distribution of susceptible material" combined with favorable climate conditions, inoculum, wind, "and complacency." With hundreds of millions of hectares planted with a susceptible wheat variety, there was plenty of combustible material. In response to the outbreak Borlaug helped establish the Global Rust Initiative (now named for him) to ensure the safety of one of the world's most important crops. The effort gathered thousands of scientists and wheat farmers from hundreds of institutions.

So far Ug99 hasn't swept around the world as scientists feared. That's a good thing. But Sarah Gurr, a plant pathologist who studies food security, stresses there has been some sensationalism around Ug99 in part because it is incredibly virulent and because so many global wheat varieties would be susceptible. There are plenty of *other* rust variants infecting wheat crops. In the 2000s a different rust strain emerged in Europe, killing crops in Sicily, Western Siberia, Denmark, Sweden, and the United Kingdom. Then there is the problem of climate. Some wheat cultivars have become more susceptible to disease as the climate changes and waves of heat descend across the fields, particularly in Europe. "Plant disease immunity changes and fungi adapt to rising temperatures," says Gurr. "We need to understand these phenomena better and hope that somewhere in wheat's genetic history is a temperature tolerant resistance gene."

Since the emergence of Ug99 and other rust strains, wheat breeders have sought—and found—resistance genes in wild relatives of wheat stored in gene banks. Still, as with other crops, the process of breeding is relatively slow. "Every time you breed a new gene into wheat or potatoes, it takes years for field trials to evaluate its usefulness," says Gurr. And now wheat breeders are on a genetic treadmill in a race against stem rusts. The fungus is exceptionally fast evolving, and traditional wheat breeders rely on single dominant resistance genes in plant genomes. But in the end, according to Gurr, you get "either 100% protection or complete disaster." And then there are the vast numbers of fungal spores. "If we could then look up in the sky, if rust is present, for every hectare in growing season there could be ten to the power of eleven spores. That's a

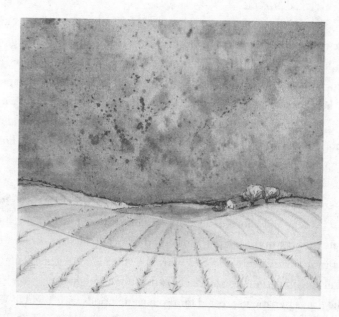

ONE HUNDRED THOUSAND MILLION SPORES

hundred thousand million spores in the sky over a field. And what they see is a palace of food." Knowing how evolvable the fungus is, the odds aren't good for our current breeding tactics, which in comparison are slow.

Some wheat strains have been genetically engineered to resist rust, but as of this writing no such engineered wheat crops are growing legally in fields. They haven't yet been approved, but they could be. "Most depressing, our number-one weapon is fungicides," which, Gurr adds, is another issue. The more popular fungicides on the market today target key enzymes that prevent fungi from functioning properly, which makes the fungicide more specific acting and potentially less toxic. But these characteristics also make it more likely that fungi will evolve resistance: perhaps a fungus acquires a slight variation on the targeted enzyme so that it remains active, or maybe a gene duplicates and so the fungus becomes able to produce more of the targeted enzyme.

One of the most common classes of agricultural fungicides are the triazoles, chemicals that target an enzyme necessary for fungi to build their cell membrane. When the membrane is compromised, a fungus fails to grow or dies. Azole-based drugs are useful therapies in humans as well. The drugs are used to treat *Aspergillus fumigatus*, among other systemic fungal infections. The soilborne fungus and its spores are everywhere and yet won't bother most of us. But for those who are immunocompromised, an *A. fumigatus* lung infection can be life-threatening. If the fungus is able to resist azole-based drugs, it is even more difficult to clear from the body and has high fatality rates (50 percent and above depending on the treatment). In 2007 scientists working in the Netherlands reported an odd finding. *A. fumigatus* cultured from several patients showed a novel kind of genetic mutation enabling azole resistance. That the fungus might evolve to resist the drugs in patients treated for prolonged periods wasn't odd; it was almost to be expected. But in this case not only were the genetics odd,

but some of the patients hadn't been exposed
to the drugs before. How could the *A. fumigatus* infecting them have evolved resistance?
The scientists surmised that the fungus had
been exposed to azoles in the environment, most
likely on farms using azole-based fungicides.

A decade later azole-resistant *A. fumigatus* was found on tulip bulbs imported from the
Netherlands to Ireland. The Netherlands produces
most of the world's tulips and more than half of the
world's flower bulbs. To protect bulbs against harmful
fungi (which incidentally does not include *A. fumigatus*),
bulbs are dipped in fungicide, and fields too may
be sprayed during the growing season. During
and after the growing season bulb waste and

ASPERGILLUS FUMIGATUS

dead tulip leaves are gathered to decompose on the compost heap
where *A. fumigatus* tends to thrive. Subsequent studies found resistant strains in the highest numbers in these mounds of compost. The
fungus was an innocent bystander that evolved resistance to a fungicide that hadn't been used to kill it. Azole-resistant *A. fumigatus* has
since been found around the globe with similar mutations, and these
resistant strains have been found not only in relation to the flower
industry but also in soil where other crops from cereals to potatoes
and strawberries had been grown. In the United States between 2006
and 2016 triazole fungicide use in agriculture quadrupled, with wheat
crops leading the way.

If fungicides are to remain a viable defense—and, Gurr suggests,
they will for the foreseeable future—we will need new fungicides
including those that act specifically against fungi (and so are less
toxic to plants and animals) and that target multiple sites. Seeking out
genetic diversity is one way for us to step away from our reliance on
toxic chemicals. We will probably never break free, but we can work

toward reduction and toward developing less toxic fungicides that are different enough from our drugs so that we aren't trading crop protection for human protection. Genetic diversity within a population or species is one thing; there is also diversity across species, strains, and varieties. How we best deploy these defenses born of DNA to safeguard our food system is a question that has yet to be resolved. A way forward will encourage broadening genetic diversity in the crops we grow and a back-to-the-future approach: cultivating a taste for different varieties of wheats, vegetables, and fruits like bananas.

* ● *

WHEN A DISEASE STRIKES A CROP LIKE THE CAVENDISH BANANA, industry's instinct is to protect or improve the crop because that is what they grow and because that is what consumers expect. This way of thinking perpetuates the practice of growing massive monocultures of cloned fruits—and we are complicit. It is time to bring both new and old varieties with different resistances and great taste into rotation and to innovate and improve cultivation techniques. Scientists working in the field say it is time to support small growers and move away from monoculture because there are generally fewer disease problems in smaller, more varied fields. In other words, diversify, diversify, diversify.

When the first round of Fusarium wilt killed off the Gros Michel banana, breeders sought to develop a disease-resistant cultivar. They didn't succeed, but there was a backup: the Cavendish variety. As a new version of Fusarium wilt spreads, the backup plan needs a backup plan. Banana breeder Rony Swennen has one. Until 2019 Swennen led the Laboratory of Tropical Crop Improvement at Katholieke Universiteit, in Leuven, Belgium, where he curated the largest collection of banana diversity in the world held under the auspices of the Bioversity International Musa Germplasm Transit Centre, or ITC. Swennen now leads the banana breeding program at the International Institute of Tropical Agriculture, a nonprofit, research-for-development institute based in Ibadan, Nigeria. Since the 1970s, when a much younger Swennen took a job at the same institute in Africa collecting specimens of banana plants from farms and forests across the continent, he has conducted research on best practices for growing the fruit. Planting a diversity of crops is one of those practices, and much of that diversity is in the ITC collection. This includes over 1,500 varieties of banana and plantain, 100 of which are edible and 40—more or less—that are sweet. There are samples from East Africa, the Democratic Republic of the Congo, Vietnam, India, and pretty much everywhere bananas grow.

Oddly, most cultivated bananas have no viable seed (consider the last banana you ate, the four tiny black dots you see are seed remnants), which also means that seeds of banana plants can't simply be collected and archived. Instead, the ITC stores clumps of banana cells that have been collected from the growing tips of banana shoots, called the meristem. Those meristems are then grown as plantlets in test tubes or frozen and stored in liquid nitrogen. When thawed, they can be used to grow tiny baby banana plantlets. At any given time in the university's basement there are thousands of test tubes topped with bright yellow plastic caps, each with a little banana plant resting in a few milliliters of substrate. The ITC sends thousands of banana samples around the world fulfilling requests from other scientists and breeders. Most of those samples are shipped to developing countries where subsistence farmers want plants that can resist Fusarium wilt or Black Sigatoka, nematodes, or climate conditions. The mass of genetic diversity at ITC is the hope for the banana's future.

Most bananas, from the Cavendish to cooking bananas, are descended from two different seed-laden wild bananas, *Musa acuminata* and *Musa balbisiana*. When cut lengthwise, *M. acuminata* is strung with rows of large black seeds like pearls, while *M. balbisiana* is a short, almost oval fruit. Our bananas are far removed from these, in appearance and taste and genetically. While both wild banana varieties have two copies of each of their eleven chromosomes and are diploid, most cultivated bananas, those we eat, are triploid: their cells contain three sets of chromosomes. Triploid plants and animals tend to be sterile. Even if those bananas *could* be bred in a traditional way, their offspring would be virtually seedless, seeds having been bred out of bananas over centuries of cultivation. Most if not all cultivated varieties of banana today are propagated as clones.

"You *can* breed any banana," says Swennen, meaning it is possible to cross one banana with another. But it is a complicated and tedious process. And then once bananas are bred, there is the challenge of finding the rare seed. The plants that are candidates for hybridization are first pollinated by hand. Once the fruits begin to turn from green to yellow, the bunch is picked and ripened. Thousands of those bananas are peeled, crushed, and sieved, all to collect genetic gold: a handful of seeds. Once collected, embryos are removed from the fertilized seeds and planted so the new plant can be evaluated for height, disease resistance, or taste. When breeders first tried to hybridize their way around Fusarium wilt, before DNA sequencing and other new technologies, it could take decades to breed a plant with desirable traits—if that was even possible. Now, according to Swennen, "We have tens of thousands of hybrids." Still, it is a slog to produce them. Swennen describes the process in numbers: given a little over 25,000 seeds collected from 13,000 bunches, a little under 4 percent, or around 800, will germinate into viable plants. Relative to other seed-bearing crops like rice or maize, this is a pittance, but each of those hybrid plants are full of possibility, from improved disease resistance to faster growth or enhanced nutrition.

Someday, in the near or far future, the bananas on the grocery shelf will look different. They will be red and short or fat; some will be starchier, some sweeter. Maybe there will even be a large variety of low-spray bananas that require fewer pesticides like there are low-spray apples. When that happens, you will know that Swennen, Kema, and others have succeeded, and that will be something to celebrate.

Bananas are just one major crop. Virtually every morsel of food we eat and grow from the soil is susceptible to fungi, bacteria, viruses, too much or too little water, changing climate, and myriad other environmental challenges. Our foods' salvation and our own is held in plants' genes.

Chapter 8

RESURRECTION

A fire tower sits on top of Mount Toby, a bump of a mountain a few miles from my home. From there, you can see south past the Holyoke Range and the Notch into Connecticut, and far to the west across the river is Mount Greylock. On the shoulders of Toby and her sister peaks, Ox Hill and Roaring Mountain, lies a patchwork of forest. Much of it is managed by the University of Massachusetts, where it was used as a Demonstration Forest—a place where students could learn about and practice forestry and conservation. There is a fragrant grove of white pines, some of which survived rust thanks in part to *Ribes* eradication efforts; old black birches; a shady hemlock forest where the path is soft with fallen needles; and a maple forest where a broken-down sugar shack has been shot full of holes. Christmas fern, hay-scented fern, bulblet fern, and dozens of other species carpet the floor with green and decorate rock crevices. And there are shrubby American chestnuts.

For years I had walked past these bushes with their long, serrated leaves and paid little attention—until one winter I noticed a chestnut. A single burr that dropped in the fall and was later revealed by the melting snow. A small chestnut managed to grow maybe twenty feet before creating that burr. Its trunk was split and cankered. About a mile away nestled in a saddle between two small peaks are the remains of a cabin; only its four-foot-wide stone hearth and chimney and a few old chestnut logs remain. The cabin had been built by a college climbing club; they used the chestnut wood that had grown in that saddle beside the hemlocks. Now only the hemlocks remain. When chestnut blight caused by the invasive fungus *Cryphonectria parasitica* struck, the university logged the trees, hoping to "clear the forest" of blight. Since then, for nearly a century, chestnut has been absent from Toby's overstory.

Shrubby stump sprouts are all that remain of the majestic tree. They grow, die from blight, and grow again from the stump. It is a *Groundhog Day* of death. When the Toby forest and points west and south all along the Appalachians lost their chestnuts, the forest ecosystem moved on. But there are hopes that someday the chestnut may rise again, taking its place among the oak and hemlock.

• ● •

ALL TRUE CHESTNUTS SHARE A COMMON ANCESTOR—SOME GREAT, great, great, great grand chestnut-like tree from which all others evolved. If you were to go back far enough before the chestnut, you might hit something resembling a mash-up of oak, chestnut, and southern beech. The first known fossils from the genus *Castanea* suggest that chestnut trees were dropping their energy-laden nuts some fifty to forty million years ago in temperate forests across the northern hemisphere from eastern Asia to North America. The ancestral tree likely hailed from Asia and spread across the continents. In time, distinct populations evolved, diverged, and adapted to their environment. By the turn of the nineteenth century the American chestnut grew fast and tall while varieties of the most commonly planted Asian chestnut species, bred as orchard trees over thousands of years, produced larger nuts on trees that were shorter in stature.

Chestnut blight hit during the heyday of the USDA agricultural-explorer program, when the agency sent adventurous young men in search of food and ornamental plants. One of those explorers was Frank Meyer, a man most comfortable searching for new foods and interesting trees in the desolate backcountry. Meyer worked under David Fairchild, Charles Marlatt's nemesis, and had once collected chestnuts from northern China, bringing them back to the States (although a year after the blight had taken hold). In 1913, when Meyer again traveled to China, he was asked if he might take a closer look at the chestnut groves there. Could he find a similar disease in Chinese chestnuts? Meyer traveled for nearly a week by mule to reach the trees, and when he did, he cabled his colleagues, writing that the trees were indeed infected, but "not a single tree could be found which had been killed entirely by the disease." He sent a fragment of fungus-infected bark and some images of the trees he had sampled to his colleagues, who found that the fungus infecting the Chinese chestnuts was "nearly identical" to the American version. Those trees had evolved a way to live with the fungus. Meyer's finding offered hope. Maybe American chestnut could be crossed with the blight-resistant tree from China. Scientists, breeders, and hobbyists have been on a quest to capture that resistance and transfer it into the American chestnut ever since. One of the earliest was the botanist Arthur Graves.

By the time Graves began breeding chestnuts in 1931, the fungus had snaked its way into billions of trees, fanning its hyphae up and around the cambium, strangling the life from every one of them. At his summer home in Hamden, Connecticut, Graves, a curator at the Brooklyn Botanic Garden, crossed Chinese (*Castanea mollissima*) and Japanese (*C. crenata*) trees with the American trees (*C. dentata*). He sought a tree that resisted blight and grew tall and strong like the American tree. At the time, there was little good science on breeding disease resistance into trees. And there were few if any resistant chestnuts left across the eastern seaboard, so breeding hybrid trees— crossing two different species—was the only option. What exactly made the Asian trees resistant and how many genes might be involved was a complete unknown.

Genetics aside, breeding chestnuts required some finesse. Separate flowers on a single tree are both male and female, but chestnuts don't self-pollinate. Male flowers lay out their pollen on long caterpillar-like clusters of flowers called catkins, and pollen grains are carried by the wind and insects to neighboring trees; female flowers are stubby, green burrs that sit in wait at the tip of a chestnut limb. Graves grew his trees in orchards and controlled pollination by sheltering maturing female flowers in paper bags. When ready to pollinate, he cut catkins from the donor males and drew them across the flower, dusting it with pollen. As the burrs matured, Graves bagged those too, protecting the growing chestnut fruits from squirrels. Over the decades as seed grew into seedling into tree, the results of each cross unfurled. Some trees resisted blight but didn't grow tall or straight. Some grew like an American tree but would then canker and die. Or a promising mix couldn't tolerate the cold winter. Graves bagged, crossed, and planted thousands of nuts for over thirty years.

CHESTNUT LEAF AND CATKIN

At least one of the thousands of seedlings born from the hundreds of trees, he was certain, would be the right mix. None were.

Years later, scientists would learn that some resistance genes might be linked with other genes. During reproduction, as genes are sorted into daughter cells, linked genes—those located close together on a chromosome—tend to "travel together," and the offspring receives all or most of them together. In humans, genes for hair and eye color are linked, brown hair with brown eyes and light hair with light eye color. Genetic linkage could explain how traits for blight resistance and branch structure seemed to travel together, which is perhaps why breeding tall, blight-resistant trees might have been a more difficult task than Graves had imagined.

While Graves matched and mated trees, contemporaries at the USDA also sought the right mix of form and function, planting tens of thousands of hybrids across more than a dozen states. One of those trees, crossed in 1946 by Russell Clapper, a USDA researcher stationed in Connecticut, showed some promise. The tree, which became known as the Clapper tree, was the product of a cross and a backcross: a Chinese American tree bred with a full American chestnut. After the initial cross, seedlings would have roughly equal numbers of genes from each parent; crossing those trees with an American chestnut gave the Clapper a more American appearance with a hefty dose of blight resistance from its Chinese grandparent. The Clapper grew for two decades, seemingly resistant. Then, on its twenty-fifth birthday, blight emerged. Five years later the tree was dead. The enormity of failure after hundreds of crosses, thousands of trees, and tens of thousands of nuts planted and grown over decades highlighted the complexities of inheritance. For frustrated breeders it seemed a good time to resign. But there will always be another dreamer.

Charles Burnham, a retired geneticist and agronomist from the University of Minnesota, knew how to shape a plant through breeding. He spent his early years with luminaries of crop-plant breeding, including Nobel Prize winner Barbara McClintock, who discovered that genes could move, changing their location within the genome, and George Beadle, who shared the prize for linking genes with gene products like enzymes. Burnham spent fifty years learning the genetic ins and outs of crop plants like maize, beans, barley, and wheat, revealing which genes did what and where they were located within the plants' genome. He knew how to breed in helpful characteristics like disease resistance and breed out whatever less-desirable characteristics might have been inherited. In the early 1980s he found a new challenge in the American chestnut. He reasoned that the past breeding projects failed because they simply didn't go far enough. Stopping after a single cross between Asian and American created a first generation, or F1, mix, which would be half one parent and half the other. Along with the resistance genes, there would be myriad others directing leaf shape and size, branch structure, and chestnut size, causing hybrids to be just that, trees with a mix of many traits. The Clapper tree with one backcross had gone a step further but still not far enough. Burnham believed characteristics of the American chestnut could be bred back into the mix through a more rigorous process of backcross breeding, a strategy originally practiced with crop plants like wheat and barley. When carried out with self-pollinating plants, which produce more genetically uniform offspring, and with plants that mature within a single growing season, backcrossing went relatively quickly. To do this with trees would mean working on a whole other timescale. Carried out in tree-time, breeding would take decades.

Like the Clapper tree, Burnham's F1 would be crossed with an American tree. This first backcross generation was called BC1. The offspring of that cross too would be crossed with another American tree. Young trees would be tested for resistance. Theoretically, by the third backcross, or BC3, wrote Burnham, the offspring would be fifteen-sixteenths American. He believed that somewhere in that one-sixteenth would be a handful of genes including resistance. When the final backcross had been made, the third generation of backcrossed chestnuts (the BC3 generation) would then be interbred for three generations. The last generation of a BC3F3 ought to be a blight-resistant, true breeding American chestnut. Burham's thirty-year plan moved chestnut genes through five generations of trees and tens of thousands of seedlings. Hundreds of helping hands would be needed for planting, measuring, and culling. Mindful of genetic diversity, the aim was to breed lines of resistant chestnuts that could endure the warmth of a Georgian summer and the snow of a Berkshire winter. Chestnut stock from the highlands or lowlands of Georgia to Tennessee, North Carolina, and Massachusetts would all become part of the program.

Burnham knew he wouldn't live to see the plan through, but an organization could carry it forward and ensure the project's success. In 1983 Burnham, along with Norman Borlaug (the agronomist who bred rust-resistant wheat) and others, founded The American Chestnut Foundation, or TAFC. Over the years geneticists, plant pathologists, flannel-clad retirees, foresters, and other citizen scientists have rallied around Burnham's vision. They are a community of scientists and chestnut romantics intent on the restoration of a tree that most have never seen in the wild and that most will never live to see flowering and fruiting in the wild. The foundation and their volunteers aren't only resurrecting a tree; they are resurrecting hope.

• ● •

WHEN BURNHAM DEVELOPED HIS PROGRAM, HE BELIEVED RESIStance came from two genes; this would make the process more complicated than it would be for a single desirable gene but nothing that couldn't be captured through crosses and backcrosses. Given that understanding, Burnham had a good plan, but it turns out that resistance genetics isn't so simple.

In the 1990s three genes were believed to be involved in resistance. We now know that several more genes help confer resistance. As in the white pine blister rust resistance breeding program, each generation of chestnut must be tested for fungal disease resistance, and part of the foundation's program requires growing out seedlings in orchards until they are old enough to be exposed to the fungus. Paul Wetzel, an ecologist, manages Smith College's MacLeish Field station and tends to an orchard full of TACF seedlings. Hundreds of Smith College students who have never seen a chestnut tree have either planted or tramped past these little trees, and the college granted TACF access to the orchard for thirty years. It contains hundreds of young chestnut trees surrounded by an electric fence to keep the deer from grazing down the saplings. The oldest were planted about nine years ago and are fruiting. All are close to the last step of the Burnham cross process; they are BC3F2s. The shorter, younger trees are in the back of the plot. Individual trees vary in their characteristics. Some trees have leaves that are longer and toothier—more deeply serrated—than others. Some trees are shorter. Others are taller. The orchard was designed with the probability of finding a handful of resistant trees, so the trees are crowded together, growing a foot apart, with about 150 plants from a single parental line in each of 9 blocks. (Newer additions to the orchard are planted farther apart to improve the chances for survival.) Interspersed are pure Chinese chestnuts; these are control trees. If shade isn't an issue, and raccoons don't dig up the seed, and deer don't nibble, then there is drought and rain and wind. Sometimes trees just die. Filling out an experimental orchard means planting many more than the targeted numbers just to get enough survivors able to grow to maturity.

When the trees are a couple of years old, they are tested for resistance. In this case the fungus is applied directly to each tree. Wetzel and volunteers from TACF make a small hole in the bark of each gangly trunk, scrape a plug of blight fungus from a petri dish, insert it into the hole, and cover the wound. They monitor and measure the cankers as they appear. Survivors get another dose of even more virulent fungus. Wetzel's goal is to identify twenty resistant trees, cull the rest to give the trees some space, and then let the survivors cross-pollinate and bear fruit that will theoretically provide blight-resistant seed for new plantings. Sadly, for all their breeding, most of the young trees in Wetzel's orchard will not have the right combination of genes. This is true of many if not all of TACF's trees that have been planted and then challenged with blight over the past decades.

Between 2002 and 2018, nearly seventy thousand BC3F3 trees—the final stage of Burnham's breeding program—were planted at TACF's research farm in Meadowview, Virginia. According to the math of genetics, if only a few genes are responsible for most of the resistance, there ought to have been about eighteen trees that were able to fend off disease like their Chinese ancestor but grow tall and fast and send out long, toothy leaves like the American tree. By 2018, though, it was clear that resistance was middling at best. The trees, when inoculated with the blight fungus, formed cankers that were less severe than those on entirely American trees but worse than those on their Chinese ancestor. The result suggested that resistance depended on more than three genes, and that whatever gene configuration conferred resistance, it had not made it into the backcross lines. The foundation is adjusting to this new reality as the breeding continues.

In 2019 a consortium of thirty-one scientists published the genome of the Chinese chestnut. The tree is the product of some forty thousand genes. According to Jared Westbrook, a quantitative geneticist and science director at The American Chestnut Foundation, resistance isn't controlled by two, three, or six genes, but many, spread across all twelve of the chestnut's chromosomes. Resistance, it turns out, is far more complex that Burnham could have imagined. Where those resistance genes are on each chromosome and what exactly they do is difficult to tell. "Like ecosystems," Westbrook explains, "genes function in networks." This means not all are directly related to an outcome like resistance but might instead *control* other genes that are turning on or off and so are affecting the outcome indirectly. The complexity has endowed the Chinese chestnuts with a lasting resistance the fungus has yet to outmaneuver. And it explains why making a resistant hybrid that grows a sheath of straight, strong, rot-resistant wood and sets out sweet American chestnuts hadn't worked as planned. Sequences from the last generation of hybrids, the BC3F3s grown at the foundation's Meadowview program, range from 65 percent American chestnut to 90 percent American chestnut, averaging out at around 83 percent. "Not exactly the 93.75 percent that Burnham was aiming for," concedes Westbrook. Nor are they nearly as resistant as Burnham had planned.

• • •

WHILE TACF WAS BREEDING, BACKCROSSING, AND CHALLENGING trees, William Powell and Charles Maynard, both at the State University of New York's School of Environmental Science and Forestry in Syracuse, took a different route to save the chestnut: genetic engineering. Their work, which has been embraced by TACF, may hold the key to resurrecting the tree.

William Powell fell in love with the idea of saving the chestnut thirty years ago. He began by studying a phenomenon called hypovirulence; a hypovirulent fungus is less able to infect and grow on a host and so less able to cause disease. In the early twentieth century as chestnut blight swept down the Appalachians, Europeans braced themselves for similar losses. Like the American trees, the native European chestnuts became infected with the blight likely carried in Asian imports. But the transformation from healthy trees to orchards of the standing dead never happened. In the 1950s an Italian plant pathologist noted that chestnuts in Genoa survived, but he couldn't explain how. A decade later a French agronomist and colleagues discovered that the difference between the American experience and theirs was the fungus. In some cases the fungus was oddly less virulent, or hypovirulent. When the agronomist spread this fungus to other infected chestnuts, cankers began to form, indicating infection, but then healed, leaving bulbous, barky scabs. Europeans tended to grow chestnuts in orchards, which helped the fungus spread, but still the trees survived. Unknown to those scientists was that a virus had infected the blight-causing fungus. Viruses infect just about every living thing on the planet from microbes to plants and animals. There are viruses that infect other viruses, and others that infect fungi and bacteria. Some kill their host. Others do no harm. This virus infected blight fungus and rendered it impotent.

In the early 1970s mycologist Sandra Anagnostakis sent for a sample of hypovirulent fungus culture from the French agronomist and began experimenting and infecting trees that already had the fungus, wondering if it could calm the disease. It worked remarkably well, converting what could have been a deadly infection into something more manageable by the tree. At the time, it seemed that the virus might be a lifeline for American trees. But the rescue never materialized. A few stands of chestnut in the United States had become naturally infected with the hypovirulent fungus, but the fungus-viral duo didn't spread as it did in Europe. Even if every tree could have been inoculated, scientists found that there were several different strains of fungus in the United States, many incompatible with the hypovirulent strain, and so the infection didn't always take. While working with the fungus, Anagnostakis and a colleague discovered something curious. Normally when the blight spores germinate and begin to feed, they release a chemical called oxalic acid or oxalate. The acid enables the fungus to break into the plants' cells and is a common strategy for plant-infecting fungi to use. The hypovirulent fungus didn't release oxalate. It lacked the chemical key.

That the fungus released a chemical which could be characterized as a form of biological warfare isn't surprising. Microbes and
animals lob all sorts of chemicals at each other. Some use chemicals
to gain or retain territory or protect themselves from being eaten.
Trees and plants run terpenes and alkaloids through their trunks
and stems. The chemicals offer some protection against attack by
insects or pathogens; terpenes are responsible for the iconic "pine"
scent. Some fungi like *Penicillium* molds release penicillin, which can
kill bacteria (we harvest it for our own benefit). When toxic chemicals are involved, there is a good chance that the target of those
chemicals has evolved some way to damp down the damage, like the
antibiotic-resistant strep and staph bacteria or antifungal-resistant
Aspergillus. Some plants attacked by oxalate-producing fungi do the
same. Bananas, strawberries, wheat, and other grains make an enzyme
called oxalate oxidase, or OxO. The enzyme has been found to hinder common fungi that use oxalate to invade plant tissues. It is the
product of a single gene called *OxO*. The American chestnut with its
coveted rot-resistant wood and long life has evolved many tricks and
chemicals to avoid the invasion, but the blight and its oxalate presented a new challenge. American chestnut *lacked* the *OxO* gene. This
was Powell's eureka moment, when his experience with biocontrol
merged with his passion for genetic engineering. What if he could
insert the *OxO* gene into the American chestnut?

This sort of application was the dream of the early bioengineers:
to make crop plants that could fend for themselves. In his book *Lords
of the Harvest* about the heady early days of genetic engineering, Dan
Charles wrote that engineers imagined themselves as green "revolu-
tionaries" who could help farmers wean themselves from toxic pesticides. In the 1980s Monsanto took the lead following the discovery
and acquisition of a gene called *Cry*—for a selective insect-killing
protein made by the *Bacillus thuringiensis* (Bt) bacterium. The protein becomes active only when ingested by certain insects and acts

from within the gut; targeted insects include larvae of moths and other insects that feed on plants. Well before Monsanto began experimenting with engineering, growers sprayed their crops with Bt as a natural pesticide; the bacterial spray is still in use today. In time Monsanto's engineers found a way to insert *Cry* genes into plants, creating Bt cotton, corn, potatoes, and others. These are now some of the most widely used genetically engineered crops and are credited with reducing insecticide use. But the revolution didn't happen as the early engineers had hoped. A subsequent project created herbicide-resistant crops. The idea was that growers could easily spray the herbicide Roundup and kill any surrounding weed or pest plant without hurting the crop. There would no longer be a need to weed or plow under crops. Most of the corn and soy now grown in the United States is "Roundup Ready." Today Bt, Roundup Ready crops, and others are planted on more than sixteen million hectares (over four hundred million acres) by millions of farmers across dozens of countries, and corporations like Monsanto have made billions of dollars from them.

As the engineered crops gained traction on the farm, organizations concerned with the health and environmental impacts of the crops have become increasingly outspoken and alarmed. As a result many countries, including much of the European Union, have banned or restricted the crops. And an in-depth review of the impacts of genetic engineering and a subsequent article by the *New York Times* in 2016 concluded that the crops have neither increased yield nor reduced overall pesticide use. Targeted insects and weeds have evolved resistance to the engineered crops, and growers have resorted to applying more pesticide. But genetic engineering is only the *tool*, and each application carries its own pros or cons or both. While crops engineered to be herbicide resistant increased herbicide use, crops like Bt cotton and corn engineered to protect against certain insects have reduced the need for certain pesticides.

Against this turbulent milieu of science, politics, and activism surrounding genetic engineering, Powell and colleagues sought to modify the chestnut, providing it with an opportunity to fend off the fungus on its own. The ultimate goal was to modify a wild tree and return it to the forest for the benefit of no one and everyone. Powell's effort has stretched across three decades of collaborators, students, and technicians. Charles Maynard, a forest geneticist, now retired, focused on learning how to grow the trees in petri dishes so that once a candidate gene was found, it could be successfully inserted into single cells. Those would eventually be grown into saplings carrying resistance in each and every cell. Some trees can be cultured from a bit of leaf, but chestnuts cannot. The forest biologist and tissue culture specialist Scott Merkle at the University of Georgia taught Maynard how to grow chestnut embryos, translucent little clusters of cells in a dish. Powell's job was to find a suitable gene and then figure out how to successfully transfer it into the American chestnut embryos. Over the years Maynard and Powell became increasingly adept at inserting genes and growing seedlings. They tried inserting several genes from the Chinese chestnut, and some provided partial resistance but others—not much. At the same time, Powell had read about plants that had been genetically modified to express the *OxO* gene. Bingo. That gene, if inserted into the American chestnut, would enable it to make its own oxalate oxidase, detoxifying the oxalic acid made by the pathogen and preventing invasion by the fungus. That was in 1997. One of the first engineered trees they made with the *OxO* gene was planted in 2006, but it didn't make enough OxO and eventually died. After ten more years of trial and error, which included testing over two dozen different genes from a Chinese chestnut tree—which either didn't pan out or were only partially effective—Maynard, Powell, and others reported the first blight-tolerant American chestnut. The key was the *OxO* gene.

In January 2020 Powell and others submitted a nearly three-hundred-page petition including appendices to the USDA for approval of the Darling 58 genetically engineered American chestnut. By then the group had tested every bit of tree from roots to shoots, including how fallen leaves might impact tadpoles to the composition of the chestnuts and the soil microbiome fostered by the transformed tree's roots. Still, the process for approval will be a slog. Powell and the TACF will have to assure the USDA that the trees cannot become plant pests, the FDA that the chestnuts are edible, and the EPA that they will not cause havoc in the environment. The petition received over three thousand comments, pro and con. In support of the tree, Caroline Pufalt, representing the Sierra Club wrote, although any genetically modified organism has uncertainties and potential risk, by "development through credible, transparent science, and guided by the precautionary principle, genetic engineering may produce an organism presenting no threat to ecosystems and which provides an environmental benefit." That is, the careful restoration of a "lost native tree." The Global Forest Coalition criticized the short-term nature of the assessment studies for the tree, writing that the studies "are completely inadequate to understand the potential impacts of a GE tree that could live hundreds of years and spread over large distances. There would be no way to call them back if something went wrong, years or decades later."

If approved, it will be years before these resistant trees set roots in ancestral soil and decades before they shed nuts enough to feed wildlife. The main value of the work, according to Powell, is to save a tree. If the genetically engineered American chestnut is approved, there will be no patent, nor will the engineered trees be sold for profit. Anyone wanting to plant and propagate the tree would be free to do so. It is a model that if successful could be used to save other trees killed off by non-native insects and disease.

• • •

THE AMERICAN CHESTNUT ENGINEERED BY POWELL AND OTHERS IS *trans*genic, meaning a gene from one organism was inserted into an entirely different organism. In the last decade the CRISPR/Cas9 system has transformed how scientists approach genetic engineering. Jennifer Doudna and Emmanuelle Charpentier won the Nobel Prize in Chemistry for developing the revolutionary technology, which is based on the immune defense of bacteria that use genome editing to disable viruses. The system relies on repetitive patterns of DNA (CRISPR) and the insertion of a protein called Cas9 that can target and cut specific regions of DNA. Doudna and Charpentier developed a way to use this "genetic scissors" to edit the genetic code of an organism. Using CRISPR, defective genes can be repaired or a gene can be turned on or off, enabling scientists to modify an organism without the insertion of foreign genes. A receptor on a plant cell used by a fungus could be turned off, preventing the fungus from gaining access. Or an immune gene might be edited so it responds more quickly. The system enables scientists to edit DNA as a writer might edit a Word document (with care taken to optimize accuracy). The technology is new enough that it hasn't yet been used to protect trees against blister rust or chestnut blight or other forest fungi. But it has shown some promise in crop plants like tomatoes, cocoa, rice, grapes, cotton, and bananas. In the near future CRISPR/Cas9 could be used to modify bananas, possibly making them better able to resist disease.

In 2017 James Dale, a biotechnologist at Queensland University in Australia, and his colleagues, reported that they had transplanted a resistance gene for Fusarium wilt from a wild banana into the Cavendish. The engineered plants carry a single gene from *Musa acuminata* ssp. *malaccensis*, a banana that evolved in Southeast Asia and, says Dale, probably coevolved with the TR4 pathogen. The resistance

gene, called *RGA2*, was highly expressed in the resistant wild plants
but hardly expressed at all in susceptible siblings. After extensive field
trials the outcome was a fungus-resistant Cavendish banana. But that
was about as far as they have gotten. "We haven't progressed these yet,
as the lines are still classified as transgenic/genetically modified," says
Dale. There is still political and public resistance to genetic engineering. But attitudes are changing, and gene editing is proving to be a
more acceptable technology.

Dale is working to enhance the expression of the *RGA2* resistance
gene in Cavendish bananas using CRISPR/Cas9. The gene exists in
the cultivar, but it is dormant. Dale hopes to turn it on. Another editing approach is to knock out "susceptibility" genes—those genes that
provide a compatible host for an opportunistic fungus. If successful,
the engineered plants—absent any foreign DNA—would likely not
be regulated as genetically modified in several countries, including
Australia, much of North and South America, and Japan. As with
Powell's chestnut, though, how the fruit will be received by growers
and consumers alike remains unknown.

• ◉ •

IF OR WHEN THE CHESTNUTS MAKE IT THROUGH THE FEDERAL
approval process, the tree will be the first plant modified for the sole
purpose of restoration and the first approved for release into the wild.
Reintroducing trees into the wild is another project altogether. Like
whitebarks, chestnuts need their space. To grow tall, they need light,
which means they need an opening. Blowdowns, fires, and lightning
strikes used to do the job. Reintroduction may require some forest management and thoughtful placement to give the trees a chance to flourish.

According to calculations by Sara Fitzsimmons, director of restoration at TACF, even if millions of resistant chestnuts are planted,
it would be centuries before anyone might wander through a mature
chestnut forest. Given the mental effort, time, and money put forth to

save a single type of tree, it is easy to ask if the expenditure is worth it for a two-thousand-year project. Why not let the forest be? It's a question that Antioch New England forest ecologist and author Tom Wessels has long considered.

One way of thinking about the loss of a species in the forest, he says, is to imagine the forest having an immune system. When a species is removed, the system becomes compromised. It can't function or resist disturbances as well, and there are always disturbances: new pests and pathogens and climate change. On the flip side, when a species goes missing, what happens to the myriad other species that depend on it for community, food, shelter? Whether we are concerned with the short term, centuries, or tens of thousands of years, coaxing the tree back, no matter how long it takes, will leave a more resilient forest for a turbulent future to come.

For some of the most troublesome fungi, once they invade, they will not "just go away." Sometimes a host species retains sufficient genetic diversity to weather the storm—at least for a while. The American chestnut did not. No one can know for certain whether a tree like the whitebark pine could survive another hundred or two hundred years and make a comeback on its own—or whether it might go the way of the chestnut had there been no human intervention.

Some of the best defenses for any species—trees, wildlife, crops— is the genetic diversity within their population. We make it difficult for species to retain diversity: we clear-cut forest, breed and plant a monocrop, clear land for our use, and cause the planet to warm.

Some species will need our help to survive. Useful genes can be transferred from one species to another and increasingly from one closely related species to another. The genetic code can be edited, hiding some genes, revealing others. As we head into the future and more species or populations continue their slide toward extinction, we will have to decide as a society how far we are willing to go to save a tree or a staple crop.

Chapter 9

CERTIFICATION

In the early to mid-1800s a potato disease caused by the fungus-like oomycete *Phytophthora infestans* swept through Europe. Wherever the disease hit, potatoes rotted in the ground. By 1845, without the starchy staple on which so many families depended for food and commerce, misery and starvation spread across Clare, Kerry, Mayo, and other Irish counties. The artist James Mahoney was sent by the *Illustrated London News* to witness the devastation. It was horrific, he wrote: "There I saw the dying, the living, and the dead, lying indiscriminately upon the same floor, without anything between them and the cold earth." Of the many factors causing the tragedy, one of the most glaring was the virtual sole reliance on potatoes for food and for trade. The tuber is not native to Europe or to North America (where blight also hit); its home is South America. Most likely potatoes crossed the ocean in the bowels of a ship full of conquistadors or traders headed back east to Europe. Over a few centuries growers bred the tuber to their liking, from a small, irregular bundle of nutrition to large, starchy white Lumpers favored by Ireland's tenant farmers. Undesirable and possibly protective traits were bred out. Monocrops of these Lumpers were planted across the land. When blight arrived, it traveled much as the early potatoes had, on a ship from the Americas sailing across the Atlantic back to Europe.

Charles Darwin, the English naturalist whose work radically changed how we understand relationships between living things and the role of natural selection and evolution, also had some ideas about the importance of geographic barriers to life and the emergence of different species. There were limits to how far and where most living things dispersed, and for species that did disperse or migrate, geographic limits often kept them apart. These limits helped to explain "life's richness." Living things that became separated in space over time evolved under the natural pressures of their environment, like the Galápagos finches and other species he studied, which varied from one island location to another. These ideas about evolution and geographic limits are relevant to disease outbreaks, including current and future fungal pandemics. Animals, plants, and microbes evolve as a community. They evolve in response to, and some even change, their environment: a tree might make more shade or provide shelter, while an abundance of algae, microbes, or small mammals might provide enough food to influence the population of other animals. Predators and their prey tend to strike a balance between life and death in which populations of both actors persist in the typical cycles of boom and bust. A rise in prey species is fodder for a growing predator population. But once those prey species are grazed down, the predators, bereft of a main food source, dwindle. Pathogens and their host sometimes behave likewise. As Darwin was figuring out the connections between living things and their surroundings from his home in the London Borough of Bromley, Europe was experiencing the consequences of mixing flora and fauna and microbes by transporting them to places they didn't belong.

When blight began circulating in Europe's potato fields, Darwin had a crop of his own in the ground—which, he noted, had turned to rot. He had collected wild potato samples from Chile, Ecuador, and elsewhere, and he sent samples to botanist friends, including his second cousin, William Darwin Fox, so they might grow and study them too. When crops across Europe began to die, scientists wondered how this could happen so pervasively and so suddenly. Darwin wondered if those potatoes he had collected from Chile might be more resilient to disease. At the time, some scientists believed that wild potatoes or potato seed (compared with the Lumpers dying from blight) collected from the tuber's native lands could help revive Europe's crops. When blight hit, Fox had been growing the Chilean potatoes. But those also died. "I fear this decides the point as to the usefulness of procuring seed from even the fountain head—the wild stock itself," wrote a defeated Fox. (Although Darwin's Chilean potatoes died, other wild potatoes have more recently been found to resist blight.)

Darwin would soon turn his attention to the mechanisms of evolution, and others would take up the work of studying infectious disease. The fact that invisible life—fungi, bacteria, and other tiny things—might cause catastrophic disease hadn't yet entered the realm of mainstream science or medicine; these were the years before Louis Pasteur and Paul Koch. The millions of microbes shimmying in a drop of pus or blood or the mycelium growing in a potato, it was believed, did not cause disease but were simply opportunists feeding on the dead and dying. So when blight struck potato crops, no one questioned that the rotting plants were covered with some sort of fungus: that was obvious. But they did argue about whether the fungus could have caused such misery. A colleague of Darwin's, the Reverend Miles Berkeley was the first to claim that blight was indeed caused by what would later be known as *Phytophthora infestans*. Though categorized initially as a fungus, it is now recognized as a water mold, an organism more closely related to plants than are fungi but that look and behave like fungi, sending out mycelia and reproducing by spores. It would be one of the first microbes to be identified as a pathogen.

The potato famine was the outcome of a dramatic convergence of human hands. We are relentless traders and travelers, and as a result we have brought species, many of which for tens of millions of years or more, lived apart—separated by the planet's oceans, islands, and mountains—crashing together. We are jumbling the world's biota, often to devastating effect. Over a century ago, recognizing the dangers of conveying pathogens and pests around the world, Charles Marlatt, the entomologist employed by the USDA, lobbied for protections. Since his time, the plant trade has exploded, expanding 500 percent in just the past half century. More than a billion plants or plant parts destined for propagation arrive at our ports every year, many of them packed into shipping containers and sent by sea on ships the length of two or more American football fields, some carrying more than twenty thousand containers. So, we may ask, how well are our plant protections working? And is it even possible to stop the next plant or crop pandemic?

• • •

MEGAN ROMBERG WORKS FOR THE USDA ANIMAL AND PLANT HEALTH Inspection Service (APHIS), and it is her job to identify fungi intercepted on imported plants or plant material at ports of entry. She can trace her job back to Flora Patterson, the mycologist who identified white pine blister rust, and to Charles Marlatt, in his crusade to prevent plant pests and pathogens from entering the country.

As one of two national mycologists (the other is John McKemy), Romberg spends her days looking through a microscope at hyphae and spores, poking into a world most of us never see. Most of the samples Romberg slips under her microscope come from ports where the US Customs and Border Protection specialists are on the front lines inspecting cut flowers, perishable fruits and vegetables, and plants for planting (seeds, corms, tubers, cuttings—whatever might be used to reproduce a specific plant). While US Customs has inspectors at every port with entry to the United States, specific locations, known as plant inspection stations, are scattered across the country. Currently, there are seventeen such stations where APHIS personnel inspect plants intended for propagation. Plants for planting get the most scrutiny. Fruits and vegetables that are eaten or may soon end up in the dumpster, a dead-end for any hitchhiking pest or pathogen, are less of a concern. Plants from one country may get more attention than plants from another, and port workers are trained to look for symptoms of infection. When an inspector suspects a fungal infection, the plant is passed on to the "mycology area identifier." There is a huge diversity of imports and a huge diversity of fungi, but there are just twenty area identifiers focused on fungi in the country. Each one, according to Romberg, sees about a thousand samples a year. In contrast, over one hundred entomology identifiers are at the ports, looking for bugs. Just about every week, Romberg finds something she has never seen before.

When Romberg sees a leaf spot, she imagines a microscopic landscape covered with star-like spores or spiral-shaped spores or long, thin ones with appendages. Because often one fungus's hyphae or mycelia tends to look much like another, spores are useful for identification. When the area identifiers are stumped, they overnight samples to Romberg or McKemy. In the past five years they have identified over one thousand different fungal taxa on plant hosts. Most samples must be processed and identified in a day or two (plants and plant parts are considered perishable, and time is money for importers), and often to accurately identify a species, it must be sporulating—that is, producing fruiting structures and spores. Sometimes a decision must be made without full confirmation of the fungal species based on the plant, where it came from, or whatever else might help identify what cannot be seen. Disease can still slip through, and when it does, within a year or two or three, some farmer will inevitably find it in her field and send a sample off to Romberg.

The most recent surprise was tar spot
fungus caused by *Phyllachora maydis* on
corn crops. In 2015 a sample arrived
from Indiana. It had popped up in
Indiana and in Illinois and soon was
infecting crops and reducing yields
throughout the Midwest. Odd-shaped,
raised, deep-brown spots the size of pin-
heads spread across infected leaves; each
spot is a fruiting body capable of releasing thousands of spores. The fungus had
come through ports in the past but was

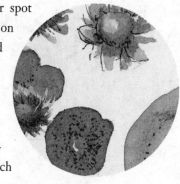

SPORES

identified, and the imports were reexported or destroyed. Four years
after it was first detected in the field, the fungus spread, infecting corn
crops in three hundred counties; heat maps show the fungus radiating away from where it was first noticed. The current plant inspection is an imperfect system, dealing with overwhelming numbers, but
still inspection teams identify hundreds of plant pathogens from the
Americas, Asia, Caribbean, Europe, and elsewhere each year; many of
them are fungi. Each pathogen detected is potentially one new blight,
smut, or rust stopped before it can ruin a crop or devastate a forest.

• • •

THERE IS NO MEGAN ROMBERG EQUIVALENT, NO NATIONAL MYCOL-
ogist or any similarly dedicated expert able to identify fungal diseases
for the wild animal trade. A novel pathogen carried by a tree frog or
other amphibian might easily pass into the country unnoticed. In 2013
when the Netherlands' fire salamander populations began to crash,
scientists identified the cause as a chytrid fungus closely related to
Bd, the frog-killer that hit Karen Lips's field sites in the early and
mid-1990s. The fungus was so virulent by the time it was discovered,
some salamander populations had been reduced to around 4 percent

FIRE SALAMANDER

of their original size. The new disease was dubbed Bsal, or *Batracho-chytrium salamandrivorans*, and like Bd, the disease is highly lethal to the naïve populations of salamanders that hadn't had the thousands or millions of years to make their peace with the fungus. The news shook herpetologists around the globe. But this time scientists could bene-fit from experience, and the outbreak was quickly linked to trade in Asian salamanders in which the fungus was thought to be endemic. A few years earlier, Asian newts imported to Europe tested positive, and the fungus had also been detected in wild populations in Vietnam. The disease was another example of the erosion of "ancient barriers to pathogen transmission." Once again we sent living organisms that had been geographically separated for millions of years crashing together.

The new incursion is frightening. It is particularly disturbing to those who work with salamanders in the United States, where the Appalachian Mountain range is a hotspot of diversity for the shy creatures. The United States is home to 25 percent of all salamander species in the world, and yet each year we import some two hundred thousand captive grown and wild-caught salamanders into the coun-try, many of them from Asia. Without inspectors on the front lines or some other means of prevention, the arrival of Bsal is almost inevita-ble. Even worse, some native salamanders are already infected with Bd. When ecologist Karen Lips witnessed the frog fiasco firsthand, it was a mystery. But Bsal is not. Lips hopes to protect the salamanders before the fungus arrives.

In 2005, when the devastation caused by Bd had become clear, representatives from organizations devoted to saving frogs, including Lips, attempted to protect whatever species they could. The group produced a report entitled "The Amphibian Conservation Action Plan." Studying, tracking, and predicting where the fungus would strike was important, they wrote, but it would not be enough. There had to be some action toward prevention. It would be "morally irresponsible to document amphibian declines and extinctions," they wrote, "without also designing and promoting a response to this global crisis." In 2009 Lips joined the faculty at the University of Maryland. There she could study amphibians in the Appalachian region, where she and others noted low rates of Bd infection in woodland salamanders. Maybe, they hoped, the salamanders had evolved some immunity against the frog fungus. The move to Maryland not only provided Lips with access to salamander hotspots; it also placed her closer to Washington, DC, the epicenter for policy-making.

When Lips arrived in Maryland, the frog decline was already on the radar of various local organizations. She was often asked to speak about her work at meetings around town, which is where she met Peter Jenkins, then director of international programs for the conservation organization Defenders of Wildlife. He had just completed work on a report entitled "Broken Screens," about the legal gaps in the animal trade. Jenkins was focused on invasive species like the red lionfish, Burmese python, and great green tree frog from Australia. As an environmental lawyer, Jenkins knew how to move policy in Washington. Lips had firsthand experience with the damages caused by a relatively unbridled animal trade industry. They would make a good team. In time the duo would pursue a wildlife disease control policy

in the manner of Charles Marlatt. And like Marlatt's travails, their
road to success would be a hard-won struggle played out over years.

Jenkins and Lips subsequently joined up with others to petition
the US Fish and Wildlife Service (FWS), the agency largely responsible for wildlife species that enter and leave the country, to help
stop the spread of Bd. Behind-the-scenes assistance came by way of
a key personality: zoologist George Rabb, an amphibian specialist.
For nearly forty years he worked with the Brookfield Zoo in Chicago
as researcher, director of education, and eventually director, and he
used his position to change the purpose of zoos from menageries for
the public to enjoy to more conservation- and education-oriented
centers. When word of disappearing amphibians first began circulating in the late 1980s, Rabb helped pull together a meeting of scientists
from around the world and developed a task force to investigate the
cause of the declines. The FWS, wrote Jenkins and Lips, needed to
"ensure traded amphibians are Bd-free based on an internationally-
recommended certification." They wrote about the Monteverde cloud
forests, the amphibian diversity in the United States, the way Bd kills,
and the US import and export of amphibians. Essentially, they were
asking the Fish and Wildlife Service to declare all amphibians injurious unless certified as disease free. The FWS deals mainly with
wildlife while the USDA Animal and Plant Health Inspection Service
protects animals raised for food, including livestock and poultry (and
many animals kept as pets).

Though there are few laws governing disease in wild animal
imports, there are plenty of laws to protect food animals from disease.
The reasons why can be seen in an earlier outbreak in Essex, England.
In February 2001, during a routine veterinary inspection of pigs at
a slaughterhouse, twenty-seven pigs appeared to be infected with
foot-and-mouth disease, a notoriously transmissible viral disease of
hooved animals. The only way to control an outbreak is to slaughter the infected herd. By March, the disease, which causes painful

blisters in the mouth and on the hooves, had infected sheep and cows, threatening economic disaster to those dealing in livestock and animal products like dairy and meat across Europe. Tens of thousands of slaughtered cattle and sheep went up in flames. In April the British government's agricultural minister claimed on television that the disease was "fully under control," even as the nation's chief veterinarian warned that the country was in the midst of a "major outbreak." Cases continued to mount as the virus popped up in France, Argentina, and Saudi Arabia. Stores ran low on meat. The usual finger-pointing and conspiracy theories ensued, while dog shows, horse races, national rugby matches, hunting activities, and larger gatherings were canceled or banned. The Republic of Ireland set up army checkpoints at all border entry points from Britain or Northern Ireland. Travel advisories warned away tourists hoping to walk the English countryside. There would be no bucolic rambles through the Cotswolds or Cornwall or along the River Cam in Cambridge that year. Wherever livestock roamed, humans could not. The general election in Britain was delayed, and protests over the slaughtering of apparently healthy animals broke out in Britain and in the Netherlands. The outbreak was estimated to have cost the United Kingdom £8 billion. US ranchers worried that it was only a matter of time before they would be culling, killing, and burning their own hooved animals.

The last major outbreak of hoof-and-mouth disease in the United States was in 1914. The virus infected 3,500 herds across the nation, and ranchers lost tens of thousands of dollars. The disease cost the United States $4.5 million (over $100 million today) to eradicate. In 1929 the virus hit again. Pigs that had been fed meat scraps from a tourist steamship sailing from Argentina became infected; 3,600 animals were slaughtered. When the outbreak in England made news in 2001, one rattled Californian dairy farmer remarked, "We're one cow away from a national disaster." President George W. Bush banned meat and meat product imports from the European Union, and the USDA-

APHIS put hundreds of animal disease experts on alert. At selected airports trained dogs sniffed returning passengers for farm smells, and passengers who arrived with soiled shoes had their footwear disinfected. Although the disease hasn't had another resurgence here, it is worrisome enough in the United Kingdom that in 2018 the agency carried out a foot-and-mouth "response exercise," a sort of "war games" for disease. And in 2020 the USDA released a sixty-four-page draft for a Foot and Mouth Response Plan with over another hundred pages of appendices. There are also disease response plans, or "Redbooks," for swine fever, avian influenza, and other diseases of livestock, all of which can easily travel and quickly cause havoc. Redbooks are the go-to guidance when a disease outbreak occurs. They provide response, control, and eradication strategies to prevent a food industry pandemic.

There's an obvious logic to the USDA's interest in monitoring and controlling animal diseases like foot and mouth and avian influenza. The goal is to keep us safe and healthy and to avoid major disruptions in our food supply and the livestock and poultry industries. But the USDA-APHIS does not have jurisdiction over wildlife. Nor does the Fish and Wildlife Service have the authority to test, upon their arrival, imported amphibians for pathogens. This is a well-known gap in the nation's regulation of disease in wildlife. The lack of regulation requiring inspection of wildlife for disease and of any coordinated monitoring or contingency plan should a novel wildlife disease emerge leaves frustratingly few options for those like Jenkins and Lips who are working to protect wildlife. But there is a historical precedent: fish.

• • •

BY THE END OF THE NINETEENTH CENTURY IT HAD BECOME CLEAR that birds, specifically European starlings and English sparrows, were a problem. The Old World birds had been imported and released where they didn't belong, in US parks and cities. The house sparrows came first, introduced by amateur ornithologist and society man Eugene Schieffelin, to control moth larvae infesting the trees around his New York City home. Other species would follow, many imported under the auspices of the American Acclimatization Society, of which Schieffelin was a prominent member. The society was modeled after the Société Zoologique d'Acclimatation, founded by French naturalist Isidore Geoffroy Saint-Hilaire in 1854. He believed that animals were malleable, imbued with the ability to adapt to new environments, and the formation of the Société provided an opportunity to test that adaptability. Yaks, llamas, and ostriches were crated and caged and sent across oceans on clipper ships and schooners; animals that survived were introduced into France or its colony in Algeria. In France some found a new home at the Jardin d'Acclimatation, the society's zoological garden in Paris. Others were intended to be domesticated. Sadly, the exotic animals were not so malleable, and many died. Saint-Hilaire's experiment failed, and he died not long after the Société's founding in 1861. Still, his work inspired other acclimatization societies, each with its own mandate, in Europe, the Americas, Australia, and elsewhere.

In England a founding member of the society imagined the countryside alive with "troops of elands gracefully galloping over the greensward and herds of koodoos." The exotic African animals, he believed, would add to the "good food of the British Isles" and beyond. In 1860 the Acclimatization Society of the British Isles introduced swans, starlings, and other birds from Europe to distant colonies. Most societies believed they were doing good work, filling the gaps that existed in nature. A society in Australia imported California quails, thrushes, and hares. Australian magpies and possums were sent to New Zealand. Black swans were sent to Calcutta. Fish, too, were moved from one stream to another and across oceans. Brown trout from England pushed out native brook trout when released into US streams. Rainbow trout native to the western

United States were dropped into waterways in Europe, Australia, and the eastern United States. Both trout species are now listed as members of the top one hundred globally invasive species by the Invasive Species Specialist Group of the International Union for Conservation of Nature.

A few decades of importing and releasing animals where they didn't belong made it increasingly obvious that tinkering with nature in this way was at the very least problematic and in some cases disastrous. On farms in the United States starlings ate the seed from newly sown crops and pulled up young plantlets. They pecked at ripening berries and peaches and other fruits. The sparrows outcompeted native bird species. This caught the attention of conservationists and of Iowa representative John Lacey. Intent on protecting wild birds, game birds, and agriculturally important birds and preventing any further releases of exotic birds and animals, on April 30, 1900, Lacey introduced legislation to Congress. "There is a compensation in the distribution of plants, birds, and animals by the God of nature," he said during the bill's introduction, adding that the disastrous introductions served as a warning to be heeded. "Let us now give an example of wise conservation of what remains of the gifts of nature."

The so-called Lacey Act was signed into law a month later. It is the nation's oldest legislation for wildlife protection. The original version of the law had two purposes: to prevent harm to agricultural interests from foreign species and to protect native species by controlling the trade and trafficking of wildlife across state lines for sport, food, and fashion. For the former purpose only wild birds and mammals were banned from import, leaving the import of fish and other vertebrates unregulated. Under the new law mongooses, fruit bats, European starlings, and English sparrows were declared "injurious wildlife" and prohibited from entering the country. Other species could be declared injurious as needed. The law also prohibited the importation of most other wild birds and wild animals *unless* allowed by special permit. Shortly after the act went into effect, changes were made to allow the import without a permit of animals known to be harmless. The number of permits required had become overwhelming for the agency in charge, which at the time was the USDA's Division of Biological Survey.

In 1948 the legislation was amended in a more profound way. The sentence "No person shall import into the United States any foreign wild animal or bird" was removed, which meant that most wild birds and mammals *would be allowed* for import unless specifically prohibited. In other words, writes Susan Jewell, injurious species coordinator at the US Fish and Wildlife Service, it was now up to the regulating authority (which had transitioned from the Department of Agriculture to the Fish and Wildlife Service; "US" was added in the 1950s) to document that a species was harmful or injurious. Soon after, they declared the prolific European rabbits and hares injurious in addition to a couple of other species, but the focus remained on birds and mammals. In 1960 Congress broadened the act's authority, and instead of applying exclusively to wild birds and wild mammals, it would now apply to all vertebrate species from fish to amphibians to reptiles along with the invertebrate groups of crustaceans and mollusks. The term "injurious" referred mainly to the potential for a species to become invasive but included reference to harm in other ways. For example, the act prohibits the importation of all members of the salmon family (which includes trout and salmon), alive or dead, and their eggs (fertilized or not) *unless* they are certified as free from certain salmonid viruses. The goal was not to prevent the spread of the fish but to prevent the introduction into the United States of the diseases they carried. And this requirement for certifiably disease-free fish is where Peter Jenkins believed there might be a chance for regulating disease-carrying frogs. If imported fish could be certified, why not amphibians?

In September 2009 the Defenders of Wildlife submitted a petition for regulation asking that the FWS do essentially the same for amphibians as they did for salmonids: list all amphibians as potentially injurious wildlife that could not be imported unless certified as Bd-free. Under the Lacey Act, they wrote in their petition, the agency directs that all amphibians "may be imported, transported, and possessed in captivity," implicitly encouraging the import and trade. "This is despite the fact that several species known to have carried Bd are regularly imported and essentially all species of amphibians can potentially act as Bd vectors or reservoirs." They asked that the clause encouraging trade not only be removed but that the act's implementing regulations include a new amendment ensuring that any animals coming or going from the United States were disease free. "Bd-infected amphibians are imported and transported interstate, without legal consequences. The Department of the Interior can stop this." After all, they wrote, the Lacey Act's salmonid certification had been relatively effective. But the two hundred or so regulated salmonid species are relatively simple to manage compared with the thousands of amphibian species to which the regulations would apply. The FWS, recognizing the gravity of the situation, sought options for controlling the Bd chytrid fungus. But with a pathogen that was already so dispersed in the United States, they questioned how effective a disease-free certification would be and so in the end didn't create one. Then came Bsal.

Despite the initial lack of success with frog protections, neither Lips nor Jenkins had given up on preventing disease through either new legislation or improved regulations. When Bsal emerged, there was one major difference from Bd: the disease hadn't yet made it into the United States. This time around, instead of asking for regulation on the whole amphibian trade, they targeted salamander imports, which was in comparison potentially more doable.

The salamander effort originated with Lips, the coauthor of a paper showing that Bsal had not yet arrived in the United States. The day before the paper was published, Jewell and Lips met with colleagues at the FWS so that Lips could present the findings. Preventing the fungus from invading the US salamander populations became a top priority for the FWS, which immediately began developing a strategy to protect the nation's salamanders coordinated by Jewell. Lips also called Jenkins, who mobilized interested parties including the Humane Society, Defenders of Wildlife, International Fund for Animal Welfare, and other amphibian interest groups; Rabb was involved in the new effort as well. Together, they approached the FWS advocating for change rather than submitting another petition. The agency encouraged Lips and her colleagues to get the Association of Zoos and Aquaria on board (or at least not against it) and, if they could, the pet industry as well. Full stakeholder buy-in would help to ensure compliance.

The group also appealed to the general public much as Charles Marlatt had done a century ago. In 2014 Lips and colleague Joseph Mendelson III wrote an op-ed published in the *New York Times*: "Stopping the Next Amphibian Apocalypse." They wrote about how for the past twenty-five years they had helplessly watched as frogs died of Bd and how Bsal moved through the animal trade. A study published a year after the editorial estimated that over a five-year period, some eight hundred thousand salamanders were imported to the United States, most of them related to species known to carry Bsal. They concluded, "We know what kind of killer we're dealing with. A global network of biologists is studying its movements. Government agencies are on the alert. Let's get it right this time."

The Fish and Wildlife Service had worked toward a solution to the Bsal problem, but unlike the USDA-APHIS, the FWS didn't have a cadre of disease certification experts able to test the numbers of animals coming into US ports. There were other bureaucratic roadblocks too, one of which was the pet trade. The pet industry didn't want all salamanders listed as injurious. So instead of requiring clean animals, in 2016 the agency banned only salamanders that posed the most risk of carrying the fungus into the United States, listing 201 salamander species as potentially injurious wildlife under the Lacey Act. After twenty years of trying to contain disease in amphibians, this was a promising step forward. Of the 201 salamanders listed, many are native to the United States and are not a large part of the animal trade, although a few listed newts (which are a kind of salamander) are known to carry disease and are traded. Relying on listing species has its issues, though. No matter how hard you look, there will always be the one you can't find that is still capable of causing disease—in this case a disease-carrying salamander or other amphibian host that has yet to be identified.

In 2018 an international group of scientists surveyed salamanders and related species for Bsal across southern China. They reported that salamanders and related amphibians in the region are essentially a reservoir of disease and that if trade continues without restriction, it "renders Bsal introduction in naïve, importing countries a near certainty." At least one of those species had not been listed in 2016 (the pet industry has voluntarily agreed to stop importing it). And of the eight hundred thousand salamanders imported into the United States reported a few years earlier, 98 percent were native to Asia.

What Lips and her colleagues want is that *all* imported amphibians be tested for disease. The current policy, however, is better than nothing at all. "Prior to the salamander rule, the US was importing 4 million amphibians a year, and none were required to be tested for any disease or quarantined or treated—that translates into 4 million chances to import a lethal pathogen." The ban has so far mitigated much of that risk for Bsal. In 2020 scientists mainly from the US Geological Survey initiated a massive program to swab and test native salamanders for disease. They swabbed and tested ten thousand animals between 2014 and 2017. None came up positive. Though the study wasn't designed to test for the efficacy of the ban, the authors suggest that such large-scale monitoring might at least nip any incursion in the bud.

• ● •

IN 2018, JUST FIVE YEARS AFTER BSAL WAS IDENTIFIED IN EUROPE and two years after the Fish and Wildlife Service's listing of some salamanders as injurious species, the European Commission issued temporary legislation (which in time would become permanent) requiring that all salamanders either traded among EU members or imported into the European Union be tested and declared free from Bsal. Pet keepers are exempt unless they engage in trade, even if they are only trading with another pet keeper. Animals that turn up positive for the disease must be treated. The treatment, says Frank Pasmans, a veterinary specialist in mycology and amphibians at the University of Ghent, isn't onerous: keeping salamanders at 25°C (77°F) for at least ten days works. "Whether traders really do that? I don't know."

Pasmans's laboratory was the first to identify Bsal as the pathogen killing salamanders in the Netherlands. Because the disease is so lethal, those who keep and breed the amphibians have a strong incentive to treat them: no one wants a reputation for spreading around diseased animals. There are always exceptions, including illegal importers from the United States or Asia. There are also gaps through which the fungus may slip. Only a portion of the animals in large shipments must be tested, and current pet owners who trade their own animals aren't yet required to certify their animals. But every little bit helps. Since the EU legislation the number of animals flowing through the commercial trade has gone down. Pasmans guesses that the process is too onerous and not worth the trouble from an economic perspective. Ideally, traded animals would be disease free, but that remains aspirational.

• ● •

FAITH CAMPBELL FEELS SIMILARLY ABOUT REGULATING PLANT imports as those who are trying to protect wildlife. A colleague of Peter Jenkins, Campbell is a political scientist by training with decades of experience as an advocate on environmental policies. She does for native plants what Jenkins, Lips, and others have been doing for amphibians: she works to protect them from invasive diseases, those inevitably missed by current quarantine and inspections. Retired from the Nature Conservancy where she was a senior policy representative, she now works alongside Jenkins with a small non-profit called the Center for Invasive Species Prevention. When she can, Campbell still meets with those who can push policy, including USDA-APHIS officials, Forest Service scientists, and members of Congress. But, she notes, "It's getting harder to find allies." As Campbell explains, generally most diseases coming into the country enter on live plants. Current plant inspectors work hard, but their job is nearly impossible. We import large quantities of plants, and APHIS inspects a small but targeted proportion of them (shipments deemed most likely to be a risk based in part on size of the shipment, the number of plants, and the plant species). Campbell isn't advocating for more inspections. Instead, she wants certification, akin to the certification required for imported salmonids; a program where foreign suppliers ensure that incoming plants are "as clean as scientists can come up with," shifting at least some of the burden outside of the United States. When plants leave the shop for our yards or the field, they ought to be as disease free as possible.

There is some precedent. APHIS already requires certification of imported geraniums. The plants can be infected with a bacterium that also infects potatoes, tomatoes, and eggplants, causing them to wilt and rot. Shipments of the plants arriving in the United States must show the plants have been tested and are free from disease. Geraniums "are grown by gazillions in South America," Campbell asserts, and most of the time the program works, though occasionally there's a slip. The bacterium showed up on plants in the spring of 2020, but it was caught early enough to stop its spread.

The larger problem for plants and animals is the lack of proactive regulation. If there was a certification program for importers, it could change the nearly impossible work of port plant inspectors. A certification program, says Campbell, would hopefully scare importers and suppliers into complying with the rules. "Another option is to just stop allowing some of these imports. Just say no." Moving disease around isn't just a problem with imported plants; the domestic trade in plants is huge. And interstate trade has been responsible for major disease outbreaks in the past and present, including an outbreak of sudden oak death in 2019.

The fungus-like *Phytophthora ramorum* began killing California's tanoaks and live oaks in the mid-1990s. The pathogen, like *Phytophthora infestans* which causes potato blight, is a water mold. The oomycete infects over one hundred different plant species in at least thirty-seven different families, many of which are effective carriers, enabling the spread of the disease. By the time the 2019 outbreak was detected, 1,600 potentially infected plants had been shipped to over a dozen states. One was Ohio, where the pathogen was found on lilacs and rhododendrons only after the plants were distributed to Walmart and Rural King stores. Homeowners who bought the plants were asked to yank the plants out by their root ball and burn or double-bag the plants and sanitize any garden tools.

One bit of hope for plants is a program called SANC (Systems Approach to Nursery Certification). Plant nurseries must work with state agencies to move nursery stock within and from their origin state. SANC is a certification program that ensures that their stock is cleaner. The program gives the growers a new way to work with state phytosanitary agencies that increases the requirements and oversight.

Ultimately, it is also a marketing strategy: nurseries opt for heightened scrutiny and regulation, and they can show their buyers that they've gone above and beyond to ensure the plants they sell are at reduced risk for transmitting diseases and pests. A pilot of the program began about a decade ago with about a dozen nurseries and is now under the auspices of the National Plant Board, a nonprofit organization of state plant pest regulatory agencies. By the end of the pilot program fourteen plant nurseries were SANC-certified. The program opened to the industry in January 2021. It is the kind of program that Campbell would like to see for imported plants. The ask, that new arrivals be certified as disease free, is not so different in theory to the requests by governments, businesses small and large, and schools that travelers, workers, and students take a COVID-19 test. Rapid diagnostics is the key to disease-free certification, and sensitive and rapid tests do exist for plant diseases, but they are not yet widely available or affordable at scale. We know now how quickly a disease spreads between lovers, neighbors, shoppers, world travelers. And we are more aware that disease can move from one species to another: a bat, pangolin, bird, or pet dog. Conservation scientists are hoping that the momentum can help move forward policies that will protect wildlife for both the sake of our own health and for the sake of nature. But if we are going to track disease in any animal, it seems we ought to do it well in humans first.

•   ●   •

LIVING THROUGH A GLOBAL PANDEMIC AND A NATIONAL FAILURE IN disease prevention, preparedness, and surveillance will impact our lives for years to come. When the SARS-CoV-2 virus took over hospitals and nursing homes, something else happened too. Diseases that had been under surveillance began to slip through the cracks; drug-resistant bacteria and fungi were given a free pass. One of them was *Candida auris*, the yeast the CDC had been tracking prior to the virus. In response to a general rise of antibiotic-resistant microbes, in

2016 the CDC developed the Antibiotic Resistance Lab Network. The network would provide states with the ability to detect and report antibiotic-resistant, disease-causing bugs—mostly bacteria—as they emerged around the country, which would in turn help to prevent the spread of resistance. *Candida* was included because of concerns with drug-resistant *Candida albicans* and others. The system is currently scaling up to include *Aspergillus* because some strains of the fungus can resist azole drugs. The system allows the agency to quickly help nursing homes, long-term care facilities, and other places where *C. auris* might pop up to contain and more effectively control the new disease. Facilities send samples to laboratories in the network that, using polymerase chain reaction (PCR) testing, can quickly identify or confirm the diagnosis. The fungus is then grown in the laboratory to determine how it will respond to antifungal drugs. Because some fungi grow slowly in the lab, the whole process can take days or weeks, but currently this remains the only way to identify drug-resistant fungi.

In those facilities where *C. auris* was found colonizing or infecting patients, institutional hygiene to prevent further spread was stepped up. Since the yeast can colonize hospital rooms and equipment and move from one patient to another on the sleeve of a gown or lab coat, staff knew to be extra careful decontaminating rooms and themselves after visiting colonized patients. Then came COVID-19. Health care workers who faced shortages in personal protective gear began reusing what they had; rooms were scrubbed for the virus but not necessarily for the fungus, which was more difficult to clear. In the race to beat the virus, screening for other infections, including *C. auris*, fell away. Not only that, but because *C. auris* favors the elderly and immunocompromised, populations also favored by the virus, the fungus blossomed. It began turning up in acute care hospitals instead of primarily long-term care facilities and in patients with no known exposure to others who might have it. Before COVID-19 the *total* case count for *C. auris* was 3,105. In 2020 alone it was 2,066. By 2021, the *annual* number of cases rose to 5,512, nearly doubling the pre-COVID total. Thousands more patients were found to have been colonized. If there is a lesson learned, it is that testing and surveillance work but do not happen often enough.

Tom Chiller, chief of the CDC's Mycologic Disease Branch, knows one of the biggest challenges is tracking disease across the nation. "For most of my life in fungi there was only *one* fungus under surveillance, Valley fever or *Coccidiosis*." The fungus, when inhaled, can make healthy people sick, which is unusual. That fungus is categorized as "reportable" in around two dozen states (each state determines its own set disease and conditions that must be reported to their public health department). Valley fever is endemic in the desert Southwest and points farther south, but in recent years it has been moving into a larger territory. Some scientists think that is because of the changing climate. The CDC tracks the disease as best it can through its National Notifiable Diseases Surveillance System: when a patient is diagnosed with any one of about 120 notifiable diseases, which include infectious diseases, potential bioterrorism agents, and sexually transmitted diseases, if a state chooses to inform the CDC, a report will land at the agency. But says Chiller, "We can't demand that a state report something."

This surveillance system is how the CDC knows that cases of *C. auris* rose during the pandemic. Chiller worries about all the other instances that aren't reported. "There are things happening in states every day, right now, that we don't know about. We really don't have a finger in the pond. We really don't have a good handle." We clearly need to rethink how we prevent and respond to pandemics in humans, plants, and wildlife.

In January 2020 a coalition of policy makers and science and tech groups in the United States launched what they called the Day One Project. The goal is to enable thinkers from across the science and technology spectrum to submit their "actionable" ideas about how to "improve the lives of all Americans." It's basically a "pitch-fest" for anyone with good ideas about solving some of our most urgent problems. The best ideas would be formatted and pushed forward by some of the most experienced policy makers in Washington, DC.

In October 2020, Karen Lips submitted a proposal titled "Improving Federal Management of Wildlife Movement and Emerging Infectious Disease." The COVID-19 pandemic, she writes, exposed vulnerabilities in the management of wildlife movement and emerging infectious diseases. The current administration has an opportunity to remedy this by creating a Task Force on the Control of Emerging Infectious Diseases to take action to better protect US citizens from animal-borne disease and to protect the nation's wildlife from disease. Lips suggests that among other actions US agencies involved with animal import and trade coordinate with each other and with international organizations to address the global movement of infectious diseases of animals. The Lacey Act could be amended, providing more power to the US Fish and Wildlife Service to identify and control wildlife disease. The Convention on International Trade in Endangered Species of Wild Fauna and Flora (CITES) could also be amended to consider disease risk. The World Organization for Animal Health could expand their disease interests beyond livestock to develop "a publicly accessible, centralized, and curated system for monitoring the global incidence and spread of wildlife pathogens in order to facilitate early detection of disease emergence and to document disease spread," Lips writes.

In short, policy makers need to take seriously the links between human, animal, and environmental health in relation to disease emergence and spread. When we send pathogens and new hosts crashing into each other and negating their geographic origins, we put ourselves at risk too. We need to take a more precautionary approach to whatever we do. All of us. We are all potential hosts to something from somewhere else. Now that we know this and understand the consequences of ignoring this truth, we can decide as individuals and as a society to take responsibility for how we operate in this world and beyond. This could mean we decide to end the trade in wild animals or animal enthusiasts (and consumers) agree to purchase only animals grown in captivity. We can decide to abide by and agree to testing protocols—especially as they become more common, able to detect a wider range of diseases, and more affordable. When we travel, we can be more thoughtful about where we've been and where we are going and if the mud on our shoes or the plant we've stuffed in our backpack might set off the next pandemic. When we have the will and the resources, we can also prevent the emergence of the next fungal pandemic.

Chapter 10

RESPONSIBILITY

In 1998 the Russian *Mir* space station developed a mold problem. The fungus was only doing what fungus does—decomposing matter—and it grew into window seals, on control panels, and around wires. Critical systems onboard the station were at risk of failing. Because the fungus had likely lived on the station for over a decade—as long as it had been inhabited—some worried that the high radiation of space could cause the fungus to mutate. In 2001 the station was brought down, much of it burning up on reentry to the planet's atmosphere and the rest scattering across the South Pacific. At the time, it was the largest spacecraft ever to fall back to Earth. Years later images surfaced from the International Space Station: a wall with three towel hooks was streaked with dark mold. Despite air filters and incessant cleaning to prevent a rehash of *Mir*, the wall looked like the back of any other bathroom door. Moisture combined with the space station's ambient temperature, which averages between a cool 21°C and 23°C (70°F and 73.4°F), created a haven for fungi. A microbial survey of the station published in 2006 found dozens of species of bacteria and fungi. More recently, scientists have suggested that it may be possible for mold to survive on the *outside* of a spacecraft.

Wherever we go, like the *Peanuts* character Pigpen who existed in a cloud of dust, we live within a cloud of microbes. We are no different than the plants and animals that carry fungus and other microbes, enabling them (with our assistance) to move over vast distances. And now we have carried terrestrial microbes beyond our planet.

Over half a century ago Joshua Lederberg, a Nobel Prize–winning microbial geneticist, worried about the consequences of moving microbes farther from home. In October 1957, while living in Australia, Lederberg watched *Sputnik* ascend into the night sky. There was no scientific evidence then that life existed elsewhere in the solar system, but Lederberg believed it could and was concerned that we could very possibly screw up first contact. Perhaps our exploration would even cause a life-form to go extinct. He had plenty of reason to worry. In his lifetime the American chestnut had become functionally extinct. Elms were dying, and humans had suffered through plenty of diseases carried from one land to another. A few years after *Sputnik*, Lederberg presented a paper at the first International Space Science Symposium in Nice, France. He warned that "history shows how the exploitation of newly found resources has enriched human experience; equally often we have seen great waste and needless misery follow from the thoughtless spread of disease and other ecological disturbances." Should some microbe travel from Earth to the Moon or Mars or beyond and happen to find favorable conditions without predators or other environmental limitations, Lederberg wondered, what would keep its population from exploding?

Our actions or inactions, he thought, might also one day reflect on the nation's scientific progress. If we explored space and left a trail of earthly DNA in the form of microbes, if DNA-based life were to exist, would we know the difference? And what of our responsibility? "Would we not deplore a heedless intrusion on other life systems?"

His contemporary, Sir Bernard Lovell, a professor of radioastronomy at the University of Manchester, wondered too about our responsibility toward extraterrestrial life. Rockets carrying earthly organisms, he wrote, might create "a moral disaster because man will have presumed the right to inject his own contaminated material into an extraterrestrial environment where organic evolution may well be in progress." Other scientists had selfish concerns: what if some novel "indigenous organisms" might one day benefit us—perhaps offering a new antibiotic? "It would be rash to predict too narrowly the ways in which undisturbed planetary surfaces, their indigenous organisms, or their molecular resources may ultimately serve human needs," wrote Lederberg.

In the early 1960s, prompted by these and other concerns, the US National Academies of Science urged the International Council of Scientific Unions to develop a set of policies to protect other planetary and interplanetary bodies. The council was a nongovernmental organization that recognized the societal benefits of scientific collaboration and cooperation across disciplines and borders, despite political, social, and economic differences (it is now part of the International Science Council). Andy Spry, a scientist at the SETI Institute and a consultant to NASA's Office of Planetary Protection, says that in the mid-1960s during the Cold War era, "the two big earthly superpowers of the age recognized that it was a good idea not to mess up the planetary science before we understood it." Space-going vessels would be decontaminated to the extent humanly possible before rocketing to the Moon or beyond. Since then, planetary protection has remained a guiding principle of international interplanetary space exploration. Before NASA's Viking landers shot off toward Mars in the 1970s, they were cleaned, swabbed for microbes, and then baked at over 110°C (230°F).

From the beginning the best technology to ensure that spacecraft were truly decontaminated was to swab, grow cultures, and count colonies, much as Louis Pasteur had done a century before. The indicator bacterium for planetary protection is the spore-forming *Bacillus subtilis*, a relatively harmless bacteria that forms tough spores. The logic is that if the spores can't survive, neither can anything else. Culturing the microbes can take days, and in this faster-paced world, that is a big downside of relying on a century-old method. Spry says that everyone wants something faster, but few alternatives have similar sensitivity and specificity of culturing despite all the advances in DNA-testing technologies. Half a century after the Viking missions, the *Perseverance* rover and its spacecraft, which launched in July 2020, would be assembled in a clean room, its durable parts baked at 150°C (302°F) and its more sensitive parts sterilized at lower temperatures or with hydrogen peroxide vapor before the whole thing was swabbed for microbes *ad nauseam*.

A few spores will always remain, and the level of cleanliness required is specific to the mission. Depending where on Mars a vessel lands, cleanliness requirements are adjusted, with life-detection missions and missions to "Special Regions" (places where our terrestrial organisms might be able to reproduce) requiring the highest standard. When the *Perseverance* took off for Mars, according to NASA policy the entire payload was allowed five hundred thousand spores and the rover itself some forty-one thousand.

Even if a microbe survives the sterilization efforts, it isn't a contamination threat unless it can survive space travel. Outside of Earth's orbit, there is close to an absolute vacuum that lacks oxygen and other matter. In a vacuum gases expand and liquids quickly boil off. There are also blasts of cosmic and solar radiation. Ionizing, ultraviolet, and heavy ions, remnants from supernovas and other events, are a constant radiation risk to space travelers and their vehicles. "Most organisms," says Spry, "don't like that." If, say, a fungus survived all that and landed on Mars, it would have other challenges. Unlike Earth, Mars lacks a significant atmosphere, something that protects Earth's surface from the Sun's more energetic ultraviolet C (UVC) rays versus the more familiar UVA and UVB rays, which are still problematic but less so. An unprotected human on the surface of Mars would get a sunburn in seconds, and direct surface exposure would be lethal to all known microorganisms, "even something relatively UV-resistant like a black mold." But this is all new territory.

In 2020 Marta Cortesão, a PhD student at the German Aerospace Center in Cologne, and colleagues published some curious results. After zapping mold spores with high levels of radiation, she and her group concluded that some fungal spores may be able to endure a trip to Mars. Years earlier, when news emerged of fungal contamination on the space stations, people started to wonder if spores could survive outside. Radiation is the number-one limitation for space travel, Cortesão says. "For human health, material, communication—everything is dependent on space radiation."

The fungus *Aspergillus niger* grows on fruits and vegetables (and can also infect lungs, although to a lesser extent than the common fungus *Aspergillus fumigatus*). Most of us would recognize it as a common black mold. Its color comes from melanin, the same molecule that protects our skin from some of the sun's ultraviolet radiation. Dark molds are known to be radiation resistant, but could they survive space travel? Cortesão and her colleagues wanted to understand how fungal spores from *A. niger* compared with bacterial spores like those of *Bacillus subtilis* and radiation-resistant extremophile bacteria like *Deinococcus radiodurans*. Cortesão says, "We exposed them to stupid amounts of radiation, testing doses up to 1,000 gray." Gray units describe the amount of energy from radiation that is absorbed per mass of tissue. "That is equivalent to many, many years in space." Humans, she says, *can* endure five gray, but "we are completely destroyed." In addition to radiation, any hitchhikers would have to contend with the vacuum of space and extreme temperatures, and yet some earthly organisms can survive. Under her experimental conditions Cortesão's fungal spores withstood high doses of ionizing radiation and could tolerate UVC better than the bacterial spores. The *Aspergillus* spores could probably survive at least the radiation for some years in real space if protected by layers and layers of cell wall components, other cells, and other physical protections.

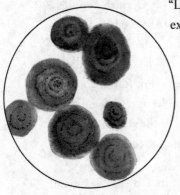

BLACK MOLD

"Layers are like mechanical shielding," she explains, "like an astronaut's suit. The melanin surrounding the spore's cell wall protects too. Current planetary protection guidelines don't really consider fungal spores. Maybe they should."

Spry isn't worried about fungal hitchhikers, although he acknowledges there are unknowns. "When you ask can an ant cross a river," he says, the answer is "no. But *do* ants cross rivers? Yes. Can a naked fungal spore survive space? Probably not." And he agrees that under layers of protection some individual microbes could possibly survive, but it would be a long shot for them to also end up somewhere warm enough and wet enough to allow them to make a home. There is one other consideration for traveling microbes: humans. From the beginning of the space program the expectation was that humans will explore. And there is the expectation that harmful contamination will *not* result as a consequence. But human exploration without some contamination is an impossibility. Sterilization techniques work well enough on ships and rovers, but humans can't be sterilized. Wherever we travel, we *will* take our terrestrial microbiome with us, even to Mars.

. ● .

IN THE EARLY DAYS OF SPACE EXPLORATION THERE WAS ANOTHER FAR
more disturbing concern than us humans contaminating new worlds.
As the world's attention focused on the once unimaginable achievements and the astronauts who put their lives at risk, some worried
about another kind of risk. What if those explorers who stepped on
the Moon brought alien microbes back to Earth when they returned?
The idea of spacecraft or astronauts contaminating Earth with alien
microbes was sci-fi frightening. "Most scientists agree that there is little chance of any life existing on the moon. But they differ widely on
the possible consequences to earth if there are lunar organisriis [*sic*]
and any of them hitch a ride with the returning astronauts," opined
an article in *Time* in June 1969. But NASA had elaborate plans for this
so-called back-contamination scenario. The agency prepared quarantines and constructed the Lunar Receiving Laboratory, a facility
designed to contain lunar microbes. The venture cost tens of millions of dollars and tens of millions more to operate. Charles Berry, a
NASA physician for the flight and retrieval of the *Apollo 11* astronauts,
recalled, "All the stuff that we went through . . . we had a swimmer in
a biological isolation garment go up to the door, but we opened that
door and threw in three biological isolation garments for the crew to
get in. But he opened that hatch, and when you open that hatch, we
had stuff come into the air, without any question about it. You know,
if it had been lunar plague, I don't know what would have happened.
I didn't believe we were going to have lunar plague, but I couldn't go
on the basis that we weren't. I mean, a lot of effort was put into trying
to prevent that from occurring."

As science historian Dagomar Degroot writes in an essay for *Aeon*
half a century later, *had* life been carried back to Earth by the Apollo
missions, despite precautions, given the limitations of the Apollo
spacecraft systems and operations, it may not have been contained.

Microbes could have infected the astronauts, or they could have been released from the Apollo capsule or released following a breach of containment. It wasn't that NASA was incompetent but that the containment task was impossible. At one point NASA had a plan to completely seal up the returning space capsule with the astronauts in it only to realize that under certain scenarios they might suffocate. The agency had to balance the safety of returning astronauts with the level of containment achieved. After exploring all the cracks in the Apollo system, Degroot concludes, "A dangerous pathogen brought to Earth from space would inevitably escape. It was only a question of when." Even so, a failure in containment wouldn't necessarily lead to a doomsday plague. As Andy Spry observes, "For that, an infection needs to occur. Also, once contained in a high reliability containment environment, sterilization can be performed to ensure a 'dangerous pathogen release' can never occur."

Half a century later the Moon is now considered devoid of life. But exploration continues to astronomical bodies that, although it is very unlikely, *may* harbor life as we know it. In the intervening decades since the original lunar exploration, planetary protection has benefited from technological advances across the sciences. In September 2021, NASA's *Perseverance* rover started drilling Mars rocks, collecting core samples in titanium tubes and sealing them closed. When these samples get returned to Earth years from now, they will be placed in another securely sealed container designed to keep everything Martian in. Eventually, before return to Earth, the container will be sterilized on the outside—to ensure it does not deliver any unknown hazards back to Earth. On return the samples will be treated as if they contain the next great plague (even if, as is most likely, they don't) and delivered to a dedicated quarantine facility designed for the highest biocontainment abilities, akin to the Level 4 laboratories approved for the study of our deadliest viruses. The hard-won samples will also be protected from terrestrial contamination. Despite the highly unlikely probability of it occurring, the effort to prevent a potential plague is huge. When compared with our lackadaisical approach to preventing the next earthly plague that is surely waiting in the wings of a bird, on the back of a frog, at the edge of the forest, and in the field, it is astonishing. That an alien plague could be brought to Earth in a sample can or by returning spacecraft is an extremely small probability event; that we will provide an opportunity for the next earthly viral, bacterial, or fungal plague to take hold is close to certainty.

• • •

WE TRAVEL, PLANT MASSIVE MONOCROPS, TRADE PLANTS AND ANImals. In the process we haven't simply opened Pandora's box; we have swung it around and shaken out the contents. We know that once a fungus settles in, there is no going back; fungi are not like viruses. As Matthew Fisher puts it, "Viruses small and simple, burn like fires." But, as eukaryotes, fungi are far more complicated. They are "very good at hedging bets and utilizing alternative hosts and cloaking themselves. Fungi are very sophisticated, biologically intelligent organisms." They will bide their time in some other species—a plant or an animal—waiting for a preferred host or lay dormant in the soil for months or years as spores. White nose syndrome continues to spread across bat populations. Cases of *Candida auris*, aspergillosis, and Valley fever are rising and will be with us for the foreseeable future. Chestnut blight is as much a threat to the American chestnut today as it was one hundred years ago. So is Dutch elm disease for elm trees. This resilience, in part, explains why some fungi have been able to drive a species into extinction. When humans first began moving animals and plants, cut forests, expanded agricultural lands, there was no thought to planetary protection. Now we know better.

Fungi *can* be capable monsters but most often only when we enable them—providing a new food source, an ever-growing population of humans with reduced immunity, or a warming environment that forces them to evolve or die. A highly virulent infection requires two things: the fungus and a susceptible host, that is, the tree or bat or frog it infects. If the frog or plant had a chance to evolve with a particular fungus, the impact could be more of an inconvenience rather than fatal. But when a potential pathogen is introduced to a new host population, it is often the host that suffers. There is another part of Pandora's story that we often forget. When she opened the box and the evils escaped, she slammed the box closed, trapping a more hesitant spirit inside—Elpis, or Hope. Some say by retaining hope, humanity was left hopeless in a world full of illness and evil, or that Hope too, particularly false hope, is an evil. Others suggest that retaining Hope in the box benefited humanity, providing us with a way to survive in a world full of evils. If we choose to believe the latter, Hope remains for our benefit. If we choose to act on our hope, we have some motivation to right our wrongs.

One of our best strategies for warding off future peril is to practice prevention. Many of the scientists whose words or research fill these pages are involved with a global initiative called One Health, which recognizes that one segment of the world's human, plant, and animal population cannot be protected from disease without considering the interconnections and the common qualities defining health in plants, animals, and the planet. The movement is reconnecting disciplines of science that over the past century have become increasingly distinct, separated by academic departments, analytical techniques, journals, language, funding, and more. At first the effort, known at the time as One Medicine, focused on connections between human medicine, veterinary medicine, and the passing of zoonotic diseases from animals to humans through direct or indirect contact. Then in 2004 the Wildlife Conservation Society organized a symposium which included health experts from around the world, that moved the conversation from primarily anthropocentric concerns of human health toward all life on Earth. One outcome was a call for a new way forward:

No one nation can reverse the patterns of habitat loss and extinction that can and do undermine the health of people and animals. Only by breaking down the barriers among agencies, individuals, specialties and sectors can we unleash the innovation and expertise needed to meet the many serious challenges to the health of people, domestic animals, and wildlife and to the integrity of ecosystems. Solving today's threats and tomorrow's problems cannot be accomplished with yesterday's approaches. We are in an era of "One World, One Health" and we must devise adaptive, forward-looking and multidisciplinary solutions to the challenges that undoubtedly lie ahead.

Thirteen years later, recognizing the universal threat from fungal pathogens, dozens of scientists published a workshop report: "One Health: Fungal Pathogens of Humans, Animals, and Plants." In it are a set of recommendations they hope will help reduce the threat of future fungal pandemics. The recommendations include better reporting and tracking of disease, finding better ways to prevent and treat disease, and developing new antifungal drugs. All of this hinges in part on a global census of *all* known fungal species in the environment and in our own human microbiome and on developing databases of fungal genomes. All will assist in better understanding outbreaks as they occur.

To this last point the scientists write that our impact on the planet has enabled the emergence of fungal pathogens. This includes species that hadn't been known to cause problems before, from *Cryphonectria parasitica* in chestnuts and *Candida auris* in humans to Pd in bats and Bd in frogs. Fungi demand our attention. If we are to preserve life on the planet as we know it, we need to take heed. Being aware of our singular actions as consumers of energy, products, and tourism is far from enough. But we can each choose to contribute in small ways.

Imagine a future where we look forward to supermarket shelves

filled not only with long, thin yellow bananas but with dozens of kinds
of bananas, some short and stubby, some blue-skinned or red. They
may be starchier or sweeter and hail from plantations large and small
that grow them as crops mixed in with papaya and other plants. Most
are labeled "sustainably grown" because the amount of pesticides and
fungicides applied have been reduced over the decades thanks to new
strains and more ecologically minded farming practices. And as major
producers move from growing just bread wheat and durum wheat to
growing other grains too, such as amaranth, teff, and buckwheat, consumers will not only embrace the different flavors and textures but
demand diversity in their grains, vegetables, and fruits.

Imagine a future with rapid diagnostics for a whole range of
microbes from viruses to fungi using DNA sequencing. Foresters who
can quickly check for disease in existing trees and plants can prevent
new outbreaks before they spread. Exporters can test their plants,
seeds, and fruits for harmful microbes and insects, and importers
can more easily swab and test biological goods themselves. What if
the trade in wild-caught animals—frogs, salamanders, fish, birds—
ceased, and instead hobbyists seek out and demand that their pets be
certified captive grown and disease free? What if the global public
recognizes that releasing captive animals into the wild, no matter
how well intentioned, often causes more harm than good? And as
world travelers, what if we consider the consequences of our movements, leaving our fruits, vegetables, and plants behind; or cleanse
our muddy shoes and boots; or submit to the occasional disease scan
when necessary?

We could consider each of these small steps as putting hope into
action. We know and care enough to believe in, act on, and support
the work of scientists and policy makers who are taking action on a
larger scale: those who work to preserve genetic diversity across species from crops to wildlife; who push for legislation to better protect
wildlife; who pursue development of more resilient plants and ani-

mals despite public pushback; or who seek to understand the genetic and ecologic underpinnings of a species' survival so that someday we might better protect them from humanity. We are the direct and indirect cause of species' extinctions and disruptions caused by fungi. We are all living in the same little boat: planet Earth. Pine trees, bats, frogs, and myriad others—we save them, we save ourselves. Acting on Hope to prevent further degradation *is* our moral obligation.

ACKNOWLEDGMENTS

My first real acquaintance with a fungus-like pathogen was when a disease called late blight spread up the East Coast in 2009, devouring late summer tomato plants that were heavy with promise. A few years later, in 2012, Mat Fisher, Sarah Gurr, and others published an article in *Nature* titled "Emerging Fungal Threats to Animal, Plant and Ecosystem Health." Other articles followed. It was as if the scientists were ringing a warning bell. They are still ringing that bell, maybe more furiously now. One goal when I began writing this book in the fall of 2019 was to amplify their warning and wake readers up to the breadth of species lost to fungal pathogens that have gone pandemic. Another goal was to remind them that this is just the beginning unless we take responsibility for our actions, which have aided these outbreaks.

Like so many of us these past few years, I met with scientists and others mostly over Zoom. We talked about fungal pandemics during a pandemic across their dining room tables and in their kitchens and living rooms. I worried that readers might not want to learn about a global fungal pandemic when we're all still processing our current pandemic. But then one scientist whose entire family had become infected with

SARS-CoV-2 reminded me that a fungal pandemic is far worse than any viral pandemic because of the nature of fungi. It's true. And so I feel some urgency to share with readers the work and words of scientists who have experienced fungal pandemics firsthand and who also have some thoughts about how to prevent future outbreaks.

I am grateful to those who gave their time to describe their work to me and then read over my chapters, answer emails, and talk some more. Their words and their work form the foundation for this book. Many also read sections of chapters to ensure accuracy. Any errors are all mine. Thank you to Giorgia Auteri, Faith Campbell, Arturo Casadevall, Marta Cortesão, Mat Fisher, Jeffrey Foster, Sarah Gurr, Peter Jenkins, Bob Keane, Gert Kema, Susan Jewell, Stuart Levitz, Marianne Moore, David Neale, Frank Pasmans, William Powell, Megan Romberg, Rony Swennen, Andy Spry, Diana Tomback, Jamie Voyles, Vance Vredenburg, Tom Wessels, Jared Westbrook, and Robert Wick.

Jon Reichard was my first interview subject when the book was just an idea for a proposal. Matt Armes provided me with insight into the world of reptile and amphibian lovers and never failed to send a detailed email response to my questions along with many useful citations. Gerald Barnes shared his memoir about the rugged and sometimes hazardous early days of the *Ribes* removal efforts and pine-tree breeding. Riley Bernard took us "batting" and allowed us to hear the local little brown bats as they whizzed by. I want to thank Brendan Jackson for so many conversations over email and for setting up a visit with the CDC, where I met Mitsuru Toda, Shawn Lockhart, Tom Chiller (virtually), and others working in the public health sector, thank you all. Karen Lips was basically "on call" and thoughtfully answered many emails. I am grateful to Luis Pocasangre, who hosted my husband and I at EARTH University in Costa Rica—what a treat. Luis provided us with an opportunity to see where our bananas come from, start to finish, along with some insight into the amount of labor involved, the history of banana plantations, and current challenges of growing our favorite fruit. Throughout the writing process, I had many conversations with Richard Sniezko via email and Zoom: he also sent links to an abundance of articles, images, and presentations. Paul Wetzel gave me a tour of the Smith College chestnut orchard and, in collaboration with the American Chestnut Foundation's MA/RI Chapter, an opportunity to experience how young chestnut trees are tested for resistance.

Many others also provided their time and expertise so that I could better understand the risk, the science, and the beauty of old growth trees, and the ongoing efforts to protect plants and animals from fungal pathogens: Ray Asselin, Beth Berkow, Thomas Bertorelli, Christine Ellis, Chad Gallinat, Andrew Gapinski, Mahmoud Ghannoum, Evan Grant, Christina Hull, Nancy Karraker, Bob Leverett, Steven Long, Joyce Longcore, Meghan Lyman, Li-Jun Ma, Colin McCormick, Douglas Minor, Nicholas Money, Brittany Moser, Taylor Perkins, John Rossow, Allison Stegner, Jeffrey Townsend, and Mark Twery. Thank you Guy Lanza for being an early reader and for your encouragement. I am also grateful the many researchers who responded to emails from a stranger titled, "Quick question for you!" which often were rarely quick or easy to answer.

I also thank the mycologists, epidemiologists, and other scientists whose work has contributed to the literature of fungus, diseases, and the living things impacted by these diseases. It's all so complicated, and the stories here are built on not just those of the scientists I've interviewed but on the work of dozens, if not hundreds, of others as well. And I am grateful to them all for working on birds and trees and frogs and water and wherever else fungi live.

Writing a book like this requires access to scientific databases and literature. I am indebted to the University of Massachusetts, Amherst, where I hold an adjunct faculty position in the Department of Environmental Conservation. The position has provided, among other things, access to the university's libraries. They are an invaluable resource. The Ronin Institute has been my academic home for over a decade; we are friends, colleagues, and cheer squad for one another, as needed.

Thank you Michelle Tessler, my agent, who helped to shape the original proposal and without whom I would not have had the opportunity to work with Melanie Tortoroli. As my editor at W. W. Norton, Melanie helped me transform words and paragraphs into a book. Her suggestions for restructuring stories and chapters and streamlining text were always on point, and her enthusiasm for the topic kept me going when I was sure no one would want to read about coming pandemics after COVID. I am also thankful that she was open to the idea of adding illustrations. I am indebted to others on the editorial and production team at W. W. Norton. Assistant editors helped move the project along: Mo Crist worked on the earliest drafts while Annabel Brazaitis gave key advice on later drafts. Publishing a book is a funny thing, authors write and editors provide invaluable guidance but getting a book (with as few errors as possible) into the hands of readers requires a whole different skill set. Thank you to the rest of the Norton team for helping to create a book that others might read: copy editor Pat Weiland ensured that my sentences (and geography, phew!) were correct and numbers accurate and Vivian Reinert proofread the manuscript. Thank you publicist Will Scarlett, marketer Steve Colca, production manager Lauren Abbate, and project editor Robert Byrne for your contributions to this effort. I would also like to thank Emily Turner from Island Press who edited my first books. A few years after my tomato plants died, over a decade ago, I drafted a proposal about killer fungi and other emergent pathogens. Emily, as always, was encouraging, and although it didn't get any further than that, I am grateful to her for those early discussions on the topic.

Just after I drafted a first chapter for the proposal, I attended the 2019 Santa Fe Science Writer's Workshop; thank you to the writing group and to lead Christie Aschwanden for reading the early draft and for their supportive comments.

Over the past three years I have come to appreciate how fortunate I am to live in a close-knit community with friends and neighbors who contributed to this project in their own way. Leigh and John Rae, our pod-buddies, got us through the last two years with cards, wine, backpacks, and skis. John read through the entire draft and provided feedback, even though he had heard plenty about deadly fungus over the past few years. Bob Strong also read early chapters about chestnut trees. It was Bob who first pointed out to me that American chestnuts still grew on Mt. Toby. Bob and Micha Archer made our trip to Costa Rica more interesting, and introduced me to Tulio who shared his experience with coffee rust. Bruce Watson, whose writing I admire, read the proposal and the very first chapter when it was a very first chapter. Julie Kumble, a good friend, created the beautiful illustrations. I could not have asked for a better experience working with an illustrator. Thank you also to Jeanne Weintraub for helping to prepare the illustrations for electronic submission. Many other neighbors, including Lee and Caroline, have heard all about fungus for several years now but are still willing to read, ask, and listen. Thank you all!

Finally, Ben, my very patient husband who had to live with this for as long as I did and was *still* happy to read the whole thing and provide encouragement and feedback despite the perils of useful critique, thank you, thank you. And finally, many thanks to the support crew: Sophie, Sam, Penny (who is also always ready to read a page, a chapter, or an entire book, and talk about the science too), and the rest of my family, including my brother-in-law Robert, who helped us to see how the illustrations might look on the page.

I could not have done any of this without you all.

NOTES

Science is a process, and new papers and citations could be added indefinitely. With that caveat the research for this book began in the summer of 2019 and concluded in the spring of 2022. As this was written during the pandemic, I relied more heavily on the literature, emails, and Zoom interviews rather than in-person visits, which made access to the literature all the more important. The notes that follow are citations to these papers: journal articles; news, magazine, and website articles; and snippets of interview material that didn't fit into the text. I hope you find them useful, and I apologize if you seek them out and hit a paywall.

INTRODUCTION

xii **a fungus called:** Jacob J. Golan and Anne Pringle, "Long-Distance Dispersal of Fungi," *Microbiology Spectrum*, 2017, 1–24.

xiii **Chemists estimate:** W. Elbert et al., "Contribution of Fungi to Primary Biogenic Aerosols in the Atmosphere: Active Discharge of Spores, Carbohydrates, and Inorganic Ions by Asco- and Basidiomycota," *Atmospheric Chemistry and Physics Discussions* 6, no. 6 (2006): 11317–55.

xiv **By some estimates:** It is difficult to find a good estimate for the numbers of species of animals, plants, or microbes. Some counts are based on "identified" and characterized species, while others, as for fungi and bacteria, may be based on both characterized species and on DNA analysis. One recent study of fungal species suggests that there may be as many as twelve million species. For more see Bing Wu et al., "Current Insights into Fungal Species Diversity and Perspective on Naming the Environmental DNA Sequences of Fungi," *Mycology* 10 (May 7, 2019): 127–40. For data on the diversity of life, see Our World in Data, "Biodiversity and Wildlife," accessed July 20, 2022, https://ourworldindata.org/biodiversity-and-wildlife#:~:text=We%20have %20identified%20and%20described,around%205%20to%2010%20million.

CHAPTER 1: EMERGENCE

3 **The emergence of a new fungal pathogen:** Centers for Disease Control and Prevention, "Fungal Diseases: Burden of Fungal Diseases in the United States," accessed February 6, 2021, https://www.cdc.gov/fungal/cdc-and-fungal/burden.html.

4 **the *New York Times* published a story:** Matt Richtel and Andrew Jacobs, "A Mysterious Infection, Spanning the Globe in a Climate of Secrecy," *New York Times*, April 6, 2019.

4 **Because the yeast resisted:** Now with enhanced cleaning and the right disinfectants, health care facilities can manage it without tearing the room apart. Brendan Jackson, personal communication, November 10, 2021.

5 **Tom Chiller is:** Tom Chiller, interview by the author, January 2020. The actual quote is, "Having a fungus like *C. auris* emerge on the scene when we still can't explain how or where it came from and it being so resistant to antifungals is a big concern."

5 **"a creature from the black lagoon":** Richtel and Jacobs, "Mysterious Infection."

5 **littered with yeasts:** Brendan R. Jackson et al., "On the Origins of a Species: What Might Explain the Rise of *Candida auris*?" *Journal of Fungi* 5, no. 3 (September 1, 2019).

6 **"wastebasket genus":** Brendan Jackson, interview by the author, November 21, 2019.

6 **yeasts reproduce asexually:** Emily Larkin et al., "The Emerging Pathogen *Candida auris*: Growth Phenotype, Virulence Factors, Activity of Antifungals, and Effect of SCY-078, a Novel Glucan Synthesis Inhibitor, on Growth Morphology and Biofilm Formation," *Antimicrobial Agents and Chemotherapy* 61, no. 5 (May 1, 2017).

6 **fantastically diverse:** Soo Chan Lee et al., "The Evolution of Sex: A Perspective from the Fungal Kingdom," *Microbiology and Molecular Biology Reviews* 74, no. 2 (June 2010): 298–340.

7 **nearly a billion patients:** Felix Bongomin et al., "Global and Multi-national Prevalence of Fungal Diseases—Estimate Precision," *Journal of Fungi* 3, no. 4 (December 1, 2017).

7 **Even before the emergence:** Chaminda J. Seneviratne and Edvaldo A. R. Rosa, "Antifungal Drug Discovery: New Theories and New Therapies" (editorial), *Frontiers in Microbiology* 7 (May 23, 2016): 728; and Bart Jan Kullberg and Maiken C. Arendrup, "Invasive Candidiasis," *New England Journal of Medicine* 373, no. 15 (October 2015): 1445–56.

7 **"disease of antibiotics":** Aya Homei and Michael Worboys, "Introduction" and "Chapter 3: Candida—a Disease of Antibiotics," in *Fungal Disease in Britain and the United States 1850–2000* (Basingstoke, UK: Palgrave Macmillan, 2013).

8 **Most fungi prefer:** Vincent A. Robert and Arturo Casadevall, "Vertebrate Endothermy Restricts Most Fungi as Potential Pathogens," *Journal of Infectious Diseases* 200, no. 10 (November 2009): 1623–26; and Monica A. Garcia-Solache and Arturo Casadevall, "Global Warming Will Bring New Fungal Diseases for Mammals," *MBio* 1, no. 1: e00061-10 (2010).

8 **the "Fungal Filter":** Arturo Casadevall, "Fungi and the Rise of Mammals," *PLoS Pathogens* 8, no. 8 (August 16, 2012).

8 **"Mammals make no sense":** Arturo Casadevall, interview by the author, November 13, 2019.

9 **"was associated with a massive":** Casadevall, interview.

9 **In 2010 Casadevall:** Garcia-Solache and Casadevall, "Global Warming."

9 **In 2019 . . . Casadevall:** Arturo Casadevall, Dimitrios P. Kontoyiannis, and Vincent Robert, "On the Emergence of *Candida auris*: Climate Change, Azoles, Swamps, and Birds," *MBio* 10, e01397-19, no. 4 (August 27, 2019).

10 **The concern, he says:** Casadevall, interview.

10 **an antifungal study:** Kaitlin Forsberg et al., "*Candida auris*: The Recent Emergence of a Multidrug-Resistant Fungal Pathogen," *Medical Mycology* 57, no. 1 (January 1, 2019): 1–12.

10 **The "ear" fungus:** Wee Gyo Lee et al., "First Three Reported Cases of Nosocomial Fungemia Caused by *Candida auris*," *Journal of Clinical Microbiology* 49, no. 9 (September 2011): 3139–42.

11 **In 2019, Brendan Jackson:** Jackson et al., "On the Origins."

11 **A subsequent paper:** Nancy A. Chow et al., "Potential Fifth Clade of *Candida auris*, Iran, 2018," *Emerging Infectious Diseases* 25, no. 9 (2019): 1780–81.

11 **quickly disabused him:** Jackson, interview; Shawn R. Lockhart et al., "Simultaneous Emergence of Multidrug-Resistant *Candida auris* on 3 Continents Confirmed by Whole-Genome Sequencing and Epidemiological Analyses," *Clinical Infectious Diseases* 64, no. 2 (2017): 134–40; and Michael A. Pfaller et al., "Twenty Years of the SENTRY Antifungal Surveillance Program: Results for Candida Species from 1997–2016," *Open Forum Infectious Diseases* 6, supplement 1 (March 15, 2019): S79–94.

11 **may have jumped:** Much has been written about this and the hypothesis that the virus was released through a laboratory leak. Richard Horton, "Offline: The Origin Story—Towards a Final Resolution?" *Lancet* 399, no. 10319 (January 1, 2022); and Edward C. Holmes et al., "The Origins of SARS-CoV-2: A Critical Review," *Cell* 184, no. 19 (2021): 4848–56. In 2022 the origin was more definitively linked to the Huanan Wholesale Food Market in Wuhan, China: Michael Worobey et al., "The Huanan Seafood Wholesale Market in Wuhan Was the Early Epicenter of the COVID-19 Pandemic," *Science* (July 26, 2022): abp8715, https://doi.org/10.1126/science.abp8715.

12 **Ebola virus:** For a remarkable study of its evolution and movement, see Philip Kiefer, "Genetic Tracking Helped Us Fight Ebola: Why Can't It Halt COVID-19?" FiveThirtyEight, ABC News, April 15, 2020, https://fivethirtyeight.com/features/genetic-tracking-helped-us-fight-ebola-why-cant-it-halt-covid-19/.

12 **a clinical oddity:** John A. Rossow et al., "A One Health Approach to Combatting *Sporothrix brasiliensis*: Narrative Review of an Emerging Zoonotic Fungal Pathogen in South America," *Journal of Fungi* 6, no. 4 (2020): 1–27.

12 **It is not unusual:** What scientists do know about the *C. auris* fungus is that it didn't travel alone. Like *Sporothrix* it likely traveled the world clinging to a human host. Nancy A. Chow et al., "Tracing the Evolutionary History and Global Expansion of *Candida auris* Using Population Genomic Analyses," *MBio* 11, no. 2 (February 16, 2022): e03364-19.

12 **All the fungus needed:** Blake M. Hanson et al., "*Candida auris* Invasive Infections during a COVID-19 Case Surge," *Antimicrobial Agents and Chemotherapy* 65, no. 10 (September 17, 2021), e01146-21.

13 **"If the T-cells don't":** Stuart M. Levitz, interview by the author, December 5, 2019.

14 **"If you had a fungal infection":** Levitz, interview.

14 **becoming more susceptible:** Arturo Casadevall, "Fungal Diseases in the 21st Century: The Near and Far Horizons," *Pathogens and Immunity* 3, no. 2 (September 25, 2018): 183.

15 **"We were seeing":** Levitz, interview.

15 **about 75 percent:** UNAIDS, "Ending AIDS: Progress towards the 90-90-90 Targets," July 20, 2017, https://www.unaids.org/en/resources/documents/2017/20170720 _Global_AIDS_update_2017; and UNAIDS, "UN AIDS Fact Sheet," accessed August, 4, 2022, https://www.unaids.org/en/resources/fact-sheet. For cropytococcal numbers see Radha Rajasingham et al., "Global Burden of Disease of HIV-Associated Cryptococcal Meningitis: An Updated Analysis," *Lancet: Infectious Diseases* 17, no. 8 (August 2017): 873–81.

16 **Levitz says physicians used to:** Levitz, interview.

16 **The echinocandins:** G. R. Thompson, T. J. Gintjee, and M. A. Donnelley, "Aspiring Antifungals: Review of Current Antifungal Pipeline Developments," *Journal of Fungi* 6, no. 1 (February 25, 2020): 1–11.

17 **existed for millennia:** K. Bhullar et al., "Antibiotic Resistance Is Prevalent in an Isolated Cave Microbiome," *PLoS One* 7, no. 4 (2012): e34953.

17 **by forming biofilms:** Mark V. Horton and Jeniel E. Nett, "*Candida auris* Infection and Biofilm Formation: Going beyond the Surface," *Current Clinical Microbiology Reports* 7 (2020): 51–56; and Larkin et al., "Emerging Pathogen *Candida auris.*"

17 **In the summer of 2021:** Meghan Lyman et al., "Transmission of Pan-Resistant and Echinocandin-Resistant *Candida auris* in Health Care Facilities—Texas and the District of Colombia, January–April 2021," *Morbidity and Mortality Weekly Report* 70 (2021): 1022–23.

18 **"Everything is in competition":** Levitz, interview.

18 **wherever it has been living:** A 2022 study found *C. auris* in apples for sale in India. Freshly picked apples did not host *C. auris,* and there is not yet any evidence linking apples for sale and the *C. auris* outbreak. Anamika Yadav et al., "*Candida auris* on Apples: Diversity and Clinical Significance," *MBio* 13, no. 2: e0051822 (March 31, 2022). *C. auris* has also been isolated from tropical coastal environments: Parth Arora et al., "Environmental Isolation of *Candida auris* from the Coastal Wetlands of Andaman Islands, India," *MBio* 12, no. 2: e03181-20 (March 1, 2021).

18 **In summer 2018:** Mitsuru Toda et al., "Notes from the Field: Multistate Coccidioidomycosis Outbreak in U.S. Residents Returning from Community Service Trips to Baja California, Mexico—July–August 2018," *Morbidity and Mortality Weekly Report* 68, no. 14 (April 12, 2019): 332–33.

19 **prior to infection:** Misuru Toda, communication with author, November 15, 2021.

19 **one model suggests:** Morgan E. Gorris et al., "Expansion of Coccidioidomycosis Endemic Regions in the United States in Response to Climate Change," *GeoHealth* 3, no. 10 (October 2019), 308–27. See the animated GIF "Climate Change to Accelerate Spread of Sometimes-Fatal Fungal Infection," accessed February 16, 2022, https://news.agu.org/press-release/climate-change-expected-to-accelerate-spread -of-sometimes-fatal-fungal-infection/.

19 **Before 1999:** Centers for Disease Control and Prevention, "Emergence of *Cryptococcus gattii*—Pacific Northwest, 2004–2010," accessed November 29, 2021, https:// www.cdc.gov/mmwr/preview/mmwrhtml/mm5928a1.htm.

19 **Casadevall and a colleague:** David M. Engelthaler and Arturo Casadevall, "On the Emergence of *Cryptococcus gattii* in the Pacific Northwest: Ballast Tanks, Tsunamis, and Black Swans," *MBio* 10:e02193-19 (October 1, 2019).

20 **the fungus colonized:** S. J. Teman et al., "Epizootiology of a *Cryptococcus gattii* Outbreak in Porpoises and Dolphins from the Salish Sea," *Diseases of Aquatic Organisms* 146 (2021): 129–43.

20 **Hospitals that had:** American Society for Microbiology, "COVID-19-Associated Mucormycosis: Triple Threat of the Pandemic," accessed January 24, 2022, https://asm .org/Articles/2021/July/COVID-19-Associated-Mucormycosis-Triple-Threat-of.

20 **Treatment often requires:** Gary M. Cox, "Mucormycosis (Zygomycosis)," UpTo-Date, accessed February 16, 2022, https://www.uptodate.com/contents/mucor mycosis-zygomycosis.

20 **the pandemic provided:** Jesil Mathew et al., "COVID-19–Associated Mucormy-cosis: Evidence-Based Critical Review of an Emerging Infection Burden during the Pandemic's Second Wave in India," *PLoS Neglected Tropical Diseases* 15, no. 11: e0009921 (November 18, 2021).

CHAPTER 2: EXTINCTION

21 **a PhD's worth:** Karen Lips, interview by the author, March 13, 2020.

22 **a lot of questions:** Lips, interview; and Karen R. Lips, "Decline of a Tropical Mon-tane Amphibian Fauna," *Conservation Biology* 12, no. 1 (February 1998): 106–17.

22 **scientists were aware:** Reviewed in Simon N. Stuart et al., "Status and Trends of Amphibian Declines and Extinctions Worldwide," *Science* 306, no. 5702 (December 3, 2004): 1783–86.

23 **Were they really "losses":** Stuart et al., "Status and Trends."

23 **no obvious explanation:** J. Alan Pounds et al., "Tests of Null Models for Amphibian Declines on a Tropical Mountain," *Conservation Biology* 11, no. 6 (1997): 1307–22.

23 **strawberry poison dart frog:** Ecological Society of America, "Strawberry Poison Frogs Feed Their Babies Poison Eggs," blog, March 20, 2014, https://www.esa.org /esablog/2014/03/20/strawberry-poison-frogs-feed-their-babies-poison-eggs/.

23 **The species went extinct:** Ed Yong, "Resurrecting the Extinct Frog with a Stomach for a Womb," *National Geographic*, March 15, 2013.

24 **ice crystals grow:** National Park Service, "Biological Miracle," last updated Decem-ber 16, 2020, https://www.nps.gov/gaar/learn/nature/wood-frog-page-2.htm.

24 **first line of protection:** Laura F. Grogan et al., "Review of the Amphibian Immune Response to Chytridiomycosis, and Future Directions," *Frontiers in Immunology* 9 (November 9, 2018): 2536.

25 **a rounded structure:** Mary L. Berbee, Timothy Y. James, and Christine Strullu-Derrien, "Early Diverging Fungi: Diversity and Impact at the Dawn of Terrestrial Life," *Annual Review of Microbiology* 71 (2017): 41–60.

25 **chitin and keratin:** One way to attract chytrid spores is to offer bits of chitin-rich shrimp shells and insect carapaces left behind after a molt. For more see Joyce E. Longcore, "Maine Chytrid Laboratory," University of Maine, accessed November 30, 2021, https://umaine.edu/chytrids/.

26 **"ask cool questions":** Lips, interview. For more see Karen R. Lips, "Witnessing Extinction in Real Time," *PLoS Biology* 16, no. 2: e2003080 (2018).

27 **a chytrid pathogenic:** Joyce E. Longcore, Allan P. Pessier, and Donald K. Nichols, "*Batrachochytrium dendrobatidis* Gen. et sp. nov., a Chytrid Pathogenic to Amphibians," *Mycologia* 91, no. 2 (1999): 219–27.

27 **some indications:** Grogan et al., "Review of the Amphibian Immune Response."

27 **a heart attack–like event:** Michael Greshko, "Ground Zero of Amphibian 'Apoca-lypse' Finally Found," *National Geographic*, May 10, 2018.

27 **"This represents the greatest documented":** Ben C. Scheele et al., "Amphibian

Fungal Panzootic Causes Catastrophic and Ongoing Loss of Biodiversity," *Science* 363, no. 6434 (March 29, 2019): 1459–63.

28 **Ox hormone induced:** Ed Yong, "How a Frog Became the First Mainstream Pregnancy Test," *Atlantic,* May 4, 2017; and Sam Kean, "The Birds, the Bees, and the Froggies," *Distillations,* August 22, 2017, https://www.sciencehistory.org/distillations /the-birds-the-bees-and-the-froggies.

28 **the Hogben test:** There was some dispute about who did what when. Lancelot Hogben, "Xenopus Test for Pregnancy," *British Medical Journal* 2, no. 4095 (July 1, 1939): 38–39.

29 **most widely distributed:** Lance van Sittert and G. John Measey, "Historical Perspectives on Global Exports and Research of African Clawed Frogs (*Xenopus laevis*)," *Transactions of the Royal Society of South Africa* 71, no. 2 (May 3, 2016): 157–66.

29 **collectors were paid:** van Sittert and Measey, "Historical Perspectives"; and D. Hey, "A Report on the Culture of the South African Clawed Frog *Xenopus laevis* (Daudin) at the Jonkershoek Inland Fish Hatchery," *Transactions of the Royal Society of South Africa* 32, no. 1 (January 1, 1949): 45–54.

29 **reports began to emerge:** US Fish and Wildlife Service, "African Clawed Frog (*Xenopus laevis*) Ecological Risk Screening Summary," September 15, 2017, https://www .fws.gov/media/ecological-risk-screening-summary-african-clawed-frog-xenopus -laevis-high-risk; and G. J. Measey et al., "Ongoing Invasions of the African Clawed Frog, *Xenopus laevis*: A Global Review," *Biological Invasions* 14, no. 11 (2012): 2255–70.

30 **thousands of dead frogs:** Vance Vredenburg, interview by the author, August 20, 2020.

30 **"We were so wrong":** Vredenburg, interview.

31 **Three of the samples:** Vance T. Vredenburg et al., "Prevalence of *Batrachochytrium dendrobatidis* in *Xenopus* Collected in Africa (1871–2000) and in California (2001– 2010)," *PLoS One* 8, no. 5 (May 15, 2013): 63791.

31 **"But it looks like":** Vredenburg, interview.

31 **study that nailed:** Simon J. O'Hanlon et al., "Recent Asian Origin of Chytrid Fungi Causing Global Amphibian Declines," *Science* 360, no. 6389 (May 11, 2018): 621–27.

31 **One study suggested:** Rhys A. Farrer et al., "Multiple Emergences of Genetically Diverse Amphibian-Infecting Chytrids Include a Globalized Hypervirulent Recombinant Lineage," *Proceedings of the National Academy of Sciences* 108, no. 46 (November 15, 2011): 18732–36; and Erica Bree Rosenblum et al., "Complex History of the Amphibian-Killing Chytrid Fungus Revealed with Genome Resequencing Data," *Proceedings of the National Academy of Sciences of the United States of America* 110, no. 23 (June 2013): 9385–90.

32 **traveled the world:** O'Hanlon et al., "Recent Asian Origin," quote on p. 3.

32 **two hundred million animals:** Peter Jenkins, Kristin Genovese, and Heidi Ruffler, "Broken Screens: The Regulation of Live Animal Imports in the United States" (Washington, DC: Defenders of Wildlife, 2007), https://defenders.org/publications /broken-screens-report.

32 **About half:** K. M. Smith et al., "Summarizing US Wildlife Trade with an Eye Toward Assessing the Risk of Infectious Disease Introduction," *EcoHealth* 14, no. 1 (2017): 29–39.

32 **In a ten-year period:** Mark Auliya et al., "Trade in Live Reptiles, Its Impact on Wild Populations, and the Role of the European Market," *Biological Conservation,* Part A, 204 (December 1, 2016): 103–19.

32 **"largest and most complex":** Smith et al., "Summarizing US Wildlife Trade."

33 **"kaleidoscope" of life:** Jonathan Kolby, "To Prevent the Next Pandemic, It's the Legal Wildlife Trade We Should Worry About," *National Geographic*, May 7, 2020.

33 **exotic snakes:** Released pets are a primary source: Florida Fish and Wildlife Conservation Commission, "Burmese Python," accessed February 16, 2022, https://myfwc.com/wildlifehabitats/profiles/reptiles/snakes/burmese-python/.

33 **"colonize so quickly":** Anika Gupta, "Invasion of the Lionfish," *Smithsonian Magazine*, May 7, 2009.

33 **"prayer animal release":** USGS, "How Did Snakehead Fish Get into the United States?" https://www.usgs.gov/faqs/how-did-snakehead-fish-get-united-states; and Kit Magellan, "Prayer Animal Release: An Understudied Pathway for Introduction of Invasive Aquatic Species," *Aquatic Ecosystem Health & Management* 22, no. 4 (October 2, 2019): 452–61.

34 **Growing up in Norfolk:** Matthew Armes, interview by the author, August 9, 2020.

34 **critically endangered:** Matthew Armes, email communication, November 15, 2021.

34 **Most exotic pet owners:** Armes, interview.

34 **"Everything you move":** Matthew Fisher, interview by the author, March 24, 2020.

35 **majority of snakes:** Jessica A. Lyons and Daniel J. D. Natusch, "Wildlife Laundering through Breeding Farms: Illegal Harvest, Population Declines and a Means of Regulating the Trade of Green Pythons (*Morelia viridis*) from Indonesia," *Biological Conservation* 144, no. 12 (December 1, 2011): 3073–81.

35 **recent studies suggest:** Jia Hao Tow, William S. Symes, and Luis Roman Carrasco, "Economic Value of Illegal Wildlife Trade Entering the USA," *PLoS One* 16, no. 10: e0258523 (October 2021) ; and Marcos A. Bezerra-Santos et al., "Legal versus Illegal Wildlife Trade: Zoonotic Disease Risks," *Trends in Parasitology* 37, no. 5 (2021): 360–61.

35 **"a functional Pangea":** Ben C. Scheele et al., "Amphibian Fungal Panzootic Causes Catastrophic and Ongoing Loss of Biodiversity," *Science* 363, no. 6434 (March 29, 2019): 1459–63. For more about illegal trade, see Wildlife Tracking Alliance, "Illegal Wildlife Trade," accessed November 29, 2021, https://wildlifetraffickingalliance.org/illegal-wildlife-trade/.

35 **American bullfrogs:** Invasive Species Compendium, "*Rana catesbeiana* (American Bullfrog)," accessed February 11, 2021, https://www.cabi.org/isc/datasheet/66618.

36 **Vredenburg's lab found:** Tiffany A. Yap et al., "Averting a North American Biodiversity Crisis: A Newly Described Pathogen Poses a Major Threat to Salamanders via Trade," *Science* 349, no. 6247 (2015): 481–82; and Yap et al., "Introduced Bullfrog Facilitates Pathogen Invasion in the Western United States," *PloS One* 13, no. 4 (April 16, 2018): e0188384.

36 **salamander-killing fungus:** Mae Cowgill et al., "Social Behavior, Community Composition, Pathogen Strain, and Host Symbionts Influence Fungal Disease Dynamics in Salamanders," *Frontiers in Veterinary Science* 8: 742288 (November 2021).

36 **Seventy-seven of those:** "AmphibiaWeb," accessed February 16, 2022, https://amphibiaweb.org/; and Yap et al., "Averting a North American Biodiversity Crisis."

37 **disappeared completely:** Elise F. Zipkin et al., "Tropical Snake Diversity Collapses after Widespread Amphibian Loss," *Science* 367, no. 6479 (2020): 814–16.

37 **"This previously unidentified":** Advancing Earth and Space Science (AGU), "Amphibian Die-Offs Worsened Malaria Outbreaks in Central America," December 2, 2020, accessed February 11, 2021, https://news.agu.org/press-release/amphibian

-die-offs-worsened-malaria-outbreaks-in-central-america/; and M. R. Springborn et al., "Amphibian Collapses Exacerbated Malaria Outbreaks in Central America," *MedRxiv*, December 9, 2020, https://doi.org/10.1101/2020.12.07.20245613.

CHAPTER 3: CATASTROPHE

39 **one plant pathologist:** Robert Wick, interview by the author, February 20, 2020.

41 **tree islands:** Diana Tomback et al., *Whitebark Pine Communities: Ecology and Restoration* (Washington, DC: Island, 2001).

41 **a varied understory:** Robert E. Keane et al., *A Range-Wide Restoration Strategy for Whitebark Pine (*Pinus albicaulis*),* Gen. Tech. Rep. RMRS-GTR-279 (Fort Collins, CO: US Department of Agriculture, Forest Service, Rocky Mountain Research Station, 2012).

41 **remarkably large seeds:** Whitebark Pine Ecosystem Foundation, "Wildlife," accessed February 16, 2022, https://whitebarkfound.org/wildlife/.

42 **high-elevation national parks:** D. F. Tomback and P. Achuff, "Blister Rust and Western Forest Biodiversity: Ecology, Values and Outlook for White Pines," *Forest Pathology* 40, no. 3–4 (August 16, 2010): 186–225.

42 **the gatekeepers:** Tomback and Achuff, "Blister Rust," 300.

42 **Scientists now worry:** Keane et al., *Range-Wide Restoration Strategy*.

43 **the first trees:** Marc D. Abrams, "Eastern White Pine Versatility in the Presettlement Forest: This Eastern Giant Exhibited Vast Ecological Breadth in the Original Forest but Has Been on the Decline with Subsequent Land-Use Changes," *BioScience* 51, no. 11 (November 1, 2001): 967–79.

43 **hundreds of years old:** Monumental Trees, "The Thickest, Tallest, and Oldest Eastern White Pines (*Pinus strobus*)," accessed December 1, 2021, https://www.monumentaltrees.com/en/trees/pinusstrobus/records/.

43 **Tree of Peace:** Haudenosaunee Confederacy, "Symbols," accessed December 1, 2021, https://www.haudenosauneeconfederacy.com/symbols/; and Indigenous Values Initiative, "Haudenosaunee Values" (see "Great Tree of Peace"), accessed December 1, 2021, https://indigenousvalues.org/haudenosaunee-values.

43 **colonial British navy:** Donald Peattie, *A Natural History of North American Trees* (San Antonio, TX: Trinity University Press, 2013), 30.

43 **"of the growth":** Peattie, *Natural History*, 32.

44 **foresters and nursery owners:** Kim E. Hummer, "History of the Origin and Dispersal of White Pine Blister Rust," *Horttechnology* 10 (2000): 515–17; P. Spaulding, United States Department of Agriculture and United States Bureau of Plant Industry, *The Blister Rust of White Pine* (Bulletin) (Washington, DC: US Government Printing Office, 1911) ; and W. V. Benedict, *History of White Pine Blister Rust Control: A Personal Account* (Washington, DC: US Department of Agriculture, Forest Service, 1981), 3.

44 **In 1909 alone:** Benedict, *History*, 4.

44 **Flora Patterson:** Appointed as mycologist in the early 1900s, Flora Patterson was the first woman in the position, which is now held by Megan Romberg. It has been held almost exclusively by women. Hannah T. Reynolds, "Flora Patterson: Ensuring That No Knowledge Is Ever Lost," in *Women in Microbiology*, ed. Rachel Whitaker and Hazel Barton (Washington, DC: American Society of Microbiology, 2018), 219–31; and "Flora W. Patterson: The First Woman Mycologist at the USDA," American

Phytopathological Society, accessed February 17, 2021, https://www.apsnet.org/ed center/apsnetfeatures/Pages/FloraPatterson.aspx.

44 **"a *Peridermium*":** G. Pierce, "White Pine Blister Rust First Report Reference," *Phytopathology* 7 (1917): 224–25.

45 **German forester:** Benedict, *History*.

45 **begins its invasion:** O. C. Maloy, "White Pine Blister Rust," *The Plant Health Instructor*, 2003, accessed July 29, 2022, https://www.apsnet.org/edcenter/disandpath/fungalbasidio/pdlessons/Pages/WhitePine.aspx.

45 **rust will invade:** Terry Tattar, "Rust Diseases," in *Diseases of Shade Trees* (Academic Press, 1989), 168–88, https://doi.org/10.1016/b978-0-12-684351-4.50017-x; and Invasive Species Compendium, "*Cronartium ribicola* (White Pine Blister Rust)," accessed February 17, 2021, https://www.cabi.org/isc/datasheet/16154.

47 **communicating with kin:** For more about the role of fungi, see Suzanne Simard, *Finding the Mother Tree* (New York: Knopf, 2021).

47 **"Russian doll–like tree":** In discussing how trees grow, Alex Shigo writes, "In effect they grow a new tree over the old one every year." Shigo, "Compartmentalization of Decay in Trees," *Scientific American*, April 1, 1985, 96.

48 **forty-two cents:** Alex Aronson, "Here's What Things Cost 100 Years Ago: Grocery Items," *Country Living*, July 30, 2020, https://www.countryliving.com/life/g3339 8396/what-things-cost-100-years-ago/?slide=3.

48 **thirty cents:** W. O. Frost, "Synopsis of Blister Rust Control Work in Maine–1932," *Blister Rust News*, vol. 17–18 (February 1933): 20.

49 **twelve million acres:** Benedict, *History*, 15.

49 **"I was first introduced to":** Gerald Barnes, personal correspondence. Barnes graciously provided me with a draft of his unpublished memoir, which I received January 2021.

50 **ended in the 1960s:** Joe Rankin, "Bad Vibes from Ribes—The Outside Story," *Northern Woodlands*, January 14, 2013, accessed February 23, 2022, https://northernwoodlands .org/outside_story/article/bad-vibes-ribes.

50 **some continued:** Isabel A. Munck et al., "Impact of White Pine Blister Rust on Resistant Cultivated *Ribes* and Neighboring Eastern White Pine in New Hampshire," *Plant Disease* 99, no. 10 (March 4, 2015): 1374–82; "Landscape: White Pine Blister Rust and *Ribes* Species," Center for Agriculture, Food, and the Environment, University of Massachusetts, Amherst, accessed December 1, 2021, https://ag.umass .edu/landscape/fact-sheets/white-pine-blister-rust-ribes-species; and Greener Grass Farm, "State Legality of Gooseberry and Currant Berry (Laws Regarding Plants in the *Ribes* Genus)," blog, February 8, 2015, https://thegreenergrassfarm .com/2015/02/08/forbidden-fruit-2-state-by-state-legality-of-gooseberry-and -currant-berry-laws-regarding-plants-in-the-ribes-genus/.

50 **banned the plants:** Munck et al., "Impact of White Pine Blister Rust"; and "Landscape: White Pine Blister Rust."

50 **There are no restrictions:** Brian W. Geils, Kim E. Hummer, and Richard S. Hunt, "White Pines, *Ribes*, and Blister Rust: A Review and Synthesis," *Forest Pathology* 40, no. 3–4 (2010): 147–85; and "State Legality of Gooseberry and Currant Berry."

50 **half the cone-producing:** Keane et al., *Range-Wide Restoration Strategy*, 31.

50 **"They are *so* dependent":** Diana Tomback, interview by the author, May 13, 2020; Tomback, "Clark's Nutcracker: Agent of Regeneration," in *Whitebark Pine Commu-*

nities, 89–104; and Tomback, "Dispersal of Whitebark Pine Seeds by Clark's Nutcracker: A Mutualism Hypothesis," *Journal of Animal Ecology* 51, no. 2 (April 22, 1982): 451–67.

52 **cache of seeds:** Tomback, interview, October 1, 2019. Tomback once estimated that nutcrackers collect tens of thousands of seeds a season. Her ballpark estimate for the number of seeds removed and hidden by a single bird was an astounding 32,000. Another estimate is that the birds cache as much as three times that amount.

52 **Both benefit:** Diana F. Tomback, "The Foraging Strategies of Clark's Nutcracker," *Living Bird* 16 (1978): 123–61.

52 **adult beetles emerge:** Ken Gibson, Sandy Kegley, and Barbara Bentz, "Forest Insect and Disease Leaflet 2: Mountain Pine Beetle," USDA Forest Service, May 2009, 1–12.

52 **favor beetle infestation:** Polly Buotte et al., "Climate Influences on Whitebark Pine Mortality from Mountain Pine Beetle in the Greater Yellowstone Ecosystem," *Ecological Applications* 26 (July 1, 2016), https://doi.org/10.1002/eap.1396.

53 **fast-growing hardwoods:** Donald Davis, "Historical Significance of American Chestnut to Appalachian Culture and Ecology," in *Restoration of American Chestnut to Forest Lands*, ed. Kim Steiner and John Carson (Asheville: North Carolina Arboretum, 2004), 53–60.

53 **nearly 30 percent:** Donald Davis, "Historical Significance of American Chestnut to Appalachian Culture and Ecology," in Steiner and Carson, eds., *Restoration of American Chestnut*, 53–60.

54 **the Zoological Society:** New York Zoological Society, "Annual Report," 1898, p. 45, https://archive.org/details/annualreportnewy31898newy/page/44/mode/2up.

55 **He built and fixed rock walls:** New York Zoological Society, "Annual Report," 1903, p. 64, https://archive.org/details/annualreportnewy81903newy/page/64/mode/2up.

55 **Merkel noticed:** Hermann Merkel, "A Deadly Fungus on the American Chestnut," in New York Zoological Society, "Annual Report," 1905, pp. 97–103, https://www.biodiversitylibrary.org/item/45339#page/105/mode/1up.

55 **a cold winter:** Susan Freinkel, *American Chestnut: The Life, Death, and Rebirth of a Perfect Tree* (Berkeley: University of California Press, 2007).

55 **all were infected:** Merkel, "Deadly Fungus," 100.

55 **misidentified it:** Freinkel, *American Chestnut*, 30.

55 **recommended fix:** Merkel, "Deadly Fungus," 101–2.

56 **the story went:** C. Campbell et al., *The Formative Years of Plant Pathology* (St. Paul, MN: APS Press, 1999), 162.

56 **A year later:** Peter Ayers, "Alexis Millardet: France's Forgotten Mycologist," *Mycologist* 18 (2004): 23–26.

56 **treated for decades:** Jay Ram Lamichhane et al., "Thirteen Decades of Antimicrobial Copper Compounds Applied in Agriculture: A Review," *Agronomy for Sustainable Development* 38, no. 3 (2018), https://doi.org/10.1007/s13593-018-0503-9.

56 **On chestnuts:** George Fiske Johnson, "The Early History of Copper Fungicides," *Agricultural History* 9, no. 2 (February 19, 1935): 67–79.

57 **Merkel asked:** D. F. Farr and A. Y. Rossman, "Fungal Databases," n.d., US National Fungus Collections, Agricultural Research Service, USDA, accessed December 17, 2019, https://nt.ars-grin.gov/fungaldatabases/.

57 **"just another timely round":** William Alphonso Murrill, *Autobiography* (Gainesville, FL: William Alphonso Murrill, 1944), 70, http://hdl.handle.net/2027/coo.31924003517681.

57 **Murrill isolated:** Freinkel, *American Chestnut*, Chapter 2, "A New Scourge."

57 *Diaporthe parasitica:* I. C. Williams, "The New Chestnut Bark Disease," *Science* 34, no. 874 (December 2, 1911): 397–400; and Murrill, *Autobiography*, https://babel .hathitrust.org/cgi/pt?id=coo.31924003517681&view=1up&seq=73.

58 **asexual spores are:** Wick, interview.

58 **"It is safe to predict":** New York Zoological Society, "Annual Report," 1906, https:// www.google.com/books/edition/Annual_Report_of_the_New_York_Zoological /GJYxAQAAMAAJ?q=chestnut&gbpv=1#f=false.

58 **fungal front:** Lawrence G. Brewer, "Ecology of Survival and Recovery from Blight in American Chestnut Trees (*Castanea dentata* (Marsh.) Borkh.) in Michigan," *Bulletin of the Torrey Botanical Club* 122, no. 1 (1995): 40–57.

59 **"wickedness":** "All Chestnuts Trees Here Are Doomed," *New York Times*, July 30, 1911.

59 **"There is no contagious disease":** Letter from the Secretary of Agriculture, May 8, 1912, Secretary Wilson, as quoted in Freinkel, *American Chestnut*, 46.

60 **around a dollar:** Sandra L. Anagnostakis, "Chestnuts and the Introduction of Chestnut Blight," Connecticut Agricultural Experiment Station, November 1997, https://portal.ct.gov/CAES/Fact-Sheets/Plant-Pathology/Chestnuts-and-the -Introduction-of-Chestnut-Blight.

61 **both had grown up:** Daniel Stone, *The Food Explorer: The True Adventures of the Globe-Trotting Botanist Who Transformed What America Eats* (New York: Dutton, 2018), 221.

61 **Marlatt was hired:** Andrew M. Liebhold and Robert L. Griffin, "The Legacy of Charles Marlatt and Efforts to Limit Plant Pest Invasions," *American Entomologist* 62, no. 4 (December 6, 2016): 218–27.

61 **Trojan horses:** Stone, *Food Explorer*, 222.

61 **During his tenure:** Liebhold and Griffin, "Legacy of Charles Marlatt."

62 **Ideally, he would have liked:** Charles Marlatt, *An Entomologist's Quest: The Story of the San Jose Scale—The Diary of a Trip around the World, 1901–1902* (Baltimore: Monumental Printing, 1953).

62 **"careless of the consequences":** Charles Marlatt, "Pests and Parasites: Why We Need a National Law to Prevent the Importation of Insect-Infested and Diseased Plants," *National Geographic*, 1911, https://www.google.com/books/edition/The _National_Geographic_Magazine/nRoRAQAAIAAJ?hl=en&gbpv=1&dq=The +entire+chestnut+timber+of+America+seems+to+be+doomed.+All+this+might +have+been+saved+with+proper+quarantine+laws&pg=PA345&printsec=frontc over, 32.

62 **"was very ominous":** Marlatt, *Entomologist's Quest*, 329.

63 **three hundred trees:** Philip J. Pauly, "The Beauty and Menace of the Japanese Cherry Trees: Conflicting Visions of American Ecological Independence," *ISIS* 87, no. 1 (1996): 51–73; and Stone, *Food Explorer*.

63 **"I found myself":** Stone, *Food Explorer*, 232.

63 **A country that:** Liebhold and Griffin, "Legacy of Charles Marlatt," 4.

63 **"wounding to Japanese Sensibilities":** "Topics of the Times," *New York Times*, January 31, 1910.

63 **profuse apologies:** Stone, *Food Explorer*, 23.

63 **an article in the widely read *National Geographic*:** Marlatt, "Pests and Parasites"; Liebhold and Griffin, "Legacy of Charles Marlatt"; and Charles Marlatt, "Farmers' Bulletin" (Washington, DC: US Government Printing Office, 1912), accessed February 10, 2021, https://tinyurl.com/2p88wvhu.

64 **In 1862:** Ironically, the vines were a potential solution to an imported fungal disease, downy mildew, which had been accidentally imported from the United States earlier. For more see George Gale, "Saving the Vine from Phylloxera," in *Wine: A Scientific Exploration,* ed. Merton Sandler and Roger Pinder (Boca Raton, FL: CRC Press, 2002), 70–91.

64 **grafting their vines:** Today the majority of wine grapes are grown this way.

64 **France began importing wine:** Kelli White, "The Devastator: *Phylloxera vastatrix* and the Remaking of the World of Wine," GuildSomm, December 30, 2017, https://www.guildsomm.com/public_content/features/articles/b/kelli-white/posts /phylloxera-vastatrix; and Javier Tello et al., "Major Outbreaks in the Nineteenth Century Shaped Grape Phylloxera Contemporary Genetic Structure in Europe," *Scientific Reports* 9, no. 1 (2019): 1–11.

65 **phytosanitary agreement:** Liebhold and Griffin, "Legacy of Charles Marlatt"; and Alan MacLeod et al., "Evolution of the International Regulation of Plant Pests and Challenges for Future Plant Health," *Food Security* 2, no. 1 (2010): 49–70.

65 **allowed the USDA:** Liebhold and Griffin, "Legacy of Charles Marlatt."

65 **"We can say":** Pauly, "Beauty and Menace," 51–73.

66 **effectively ended:** Liebhold and Griffin, "Legacy of Charles Marlatt."

67 **A. psidii hopscotched:** In Hawaii it infected 'ōhi'a, causing what was called at the time 'ōhi'a rust. The trees, the most numerous on the islands, are cultural touchstones and foundational to the island ecology. Their loss, for any reason, would be utterly devastating. Rust damage was modest, but around 2013 the 'ōhi'a began dying in large numbers. An entirely different fungus, an aggressive member of the *Ceratocystis* genus, had arrived. Hundreds of thousands of 'ōhi'a on the island of Hawaii have been killed and Hawaiians are desperately trying to prevent its spread to the other islands. Lloyd Loope, "Guidance Document for Rapid 'Ohi'a Death," December 2016, https://www.fs.fed.us/psw/publications/hughes/psw _2016_hughes006_loope.pdf.

67 **No one knows:** Roderick J. Fensham et al., "Imminent Extinction of Australian Myrtaceae by Fungal Disease," *Trends in Ecology & Evolution* 35, no. 7 (July 1, 2020): 554–57.

67 **in Australia:** "*Austropuccinia psidii* (Myrtle Rust)," Invasive Species Compendium, accessed April 25, 2022, https://www.cabi.org/isc/datasheet/45846; and Plant Health Australia, "Threat Specific Contingency Plan: Guava (Eucalyptus) Rust *Puccinia psidii,*" March 2009, www.planthealthaustralia.com.au/wp-content/uploads/2013/03 /Guava-or-Eucalyptus-rust-CP-2009.pdf.

67 **A rapid DNA test:** M. Glen et al., "*Puccinia psidii*: A Threat to the Australian Environment and Economy—A Review," *Australasian Plant Pathology* 36, no. 1 (2007): 1–16; and Inez C. Tommerup et al., "Guava Rust in Brazil: A Threat to Eucalyptus and Other Myrtaceae," *New Zealand Journal of Forestry Science* 33 (2003): 420–28.

67 **contingency plans:** Angus J. Carnegie and Geoff S. Pegg, "Lessons from the Incursion of Myrtle Rust in Australia," *Annual Review of Phytopathology* 56 (August 25, 2018): 457–78.

67 **16,000 nursery plants:** Glen et al., "*Puccinia psidii.*"

68 **"is on a steep trajectory":** Fensham et al., "Imminent Extinction of Australian Myrtaceae."

68 **in New Zealand:** Carnegie and Pegg, "Lessons."

CHAPTER 4: SUSTENANCE

69 **Per person in the United States:** Economic Research Service, USDA, "Apples and Oranges Are the Top U.S. Fruit Choices," updated August 25, 2021, accessed December 2, 2021, https://www.ers.usda.gov/data-products/chart-gallery/gallery /chart-detail/?chartId=58322.

69 **After maize:** International Plant Biotechnology Outreach, "Bananas: The Green Gold of the South" (Ghent, Belgium: IPBO, 2021), https://ipbo.sites.vib.be/sites/ipbo .sites.vib.be/files/2021-01/Bananas the green gold of the South.pdf.

69 **most are grown:** Food and Agriculture Organization of the United Nations (FAO), "Banana Market Review 2019," February 2020, 7, http://www.fao.org/3/cb0168en /cb0168en.pdf.

70 **over seventy million people:** J. G. Adheka et al., "Plantain Diversity in the Democratic Republic of Congo and Future Prospects," in *Acta Horticulturae* 1225 (2018): 261–68; and International Plant Biotechnology Outreach, "Bananas."

70 **four hundred million people:** Sabine Altendorf, "Banana Fusarium Wilt Tropical Race 4: A Mounting Threat to Global Banana Markets?" FAO Food Outlook, November 2019, https://www.fao.org/3/ca6911en/CA6911EN_TR4EN.pdf.

70 **$40 billion global industry:** Altendorf, "Banana Fusarium Wilt, " 15.

71 **"bananas were everywhere":** Luis Pocasangre, interview by the author, November 5, 2019.

71 **Rowe worked:** Dan Koeppel, *Banana: The Fate of the Fruit That Changed the World* (New York: Plume, 2008), 158. Rowe also recognized, as an American, the importance of social responsibility. He was known for providing food, financial assistance, and advice to those who asked. When Rowe died in 2001, a writer for *El Tiempo*, the Honduran daily newspaper, wrote, "We have lost the best American who ever came to Honduras," according to Anne Vezina, "Tribute to Phil Rowe," ProMusa, July 29, 2013, https://www.promusa.org/blogpost307-Tribute-to-Phil-Rowe.

71 **study of children:** Berna van Wendel de Joode et al., "Indigenous Children Living Nearby Plantations with Chlorpyrifos-Treated Bags Have Elevated 3,5,6-Trichloro-2-Pyridinol (TCPy) Urinary Concentrations," *Environmental Research* 117 (August 2012): 17–26.

73 **The biological treatment:** Pocasangre, interview.

73 **"A real bananara":** Pocasangre, interview.

74 **Race-1 strains:** Gert Kema, interview by the author, September 15, 2020; and N. Maryani et al., "Phylogeny and Genetic Diversity of the Banana Fusarium Wilt Pathogen *Fusarium oxysporum* f. sp. *cubense* in the Indonesian Centre of Origin," *Studies in Mycology* 92 (March 1, 2019): 155–94.

74 **cultivar's popular history:** Koeppel, *Banana*, 32–33.

74 **dark history:** John Soluri, *Banana Cultures* (Austin: University of Texas Press, 2005); and Koeppel, *Banana*.

74 **By 1913:** Soluri, "Accounting for Taste," 390.

75 **flooding suffocated:** Gert Kema, "The Ongoing Pandemic of Tropical Race 4 Threatens Global Banana Production," Open Plant Pathology, YouTube, May 1, 2020, https://www.youtube.com/watch?v=t9DARCO0wE8.

75 **draining millions:** Soluri, "Accounting for Taste," 396.

76 **William Cavendish:** Koeppel, *Banana*, 138.

76 **eponymous jingle:** Koeppel, *Banana*, 117.

76 *Fusarium odoratissimum:* Dirk Albert Balmer et al., "Editorial: Fusarium Wilt of Banana, a Recurring Threat to Global Banana Production," *Frontiers in Plant Science* (January 11, 2021): 628888, https://doi.org/10.3389/fpls.2020.628888.

77 **passageways clog:** Miguel Dita et al., "Fusarium Wilt of Banana: Current Knowledge on Epidemiology and Research Needs toward Sustainable Disease Management," *Frontiers in Plant Science* (October 19, 2018), https://doi.org/10.3389/fpls.2018 .01468; and ProMusa, "*Fusarium oxysporum* f. sp. *cubense*," n.d., https://www.promusa .org/Fusarium+oxysporum+f.+sp.+cubense.

77 **no way to remove:** Food and Agriculture Organization of the United Nations, "Preventing the Spread and Introduction of Banana Fusarium Wilt Disease Tropical Race 4 (TR4): Guide for Travelers," Rome, 2020, http://www.fao.org/3/ca7590en/ca 7590en.pdf; and Kema, "Ongoing Pandemic of Tropical Race 4," 8.

77 **A story about the fungus:** Jacopo Prisco, "Why Bananas as We Know Them Might Go Extinct (Again)," CNN, January 8, 2016, https://www.cnn.com/2015/07/22/ africa/banana-panama-disease/index.html; Dan Koeppel, "Yes We Will Have No Bananas," *New York Times*, June 18, 2008; Mike Reed, "We Have No Bananas," *The New Yorker*, January 10, 2010.

77 **Indonesia:** Marcel Maymon et al., "The Origin and Current Situation of *Fusarium oxysporum* f. sp. *cubense*' Tropical Race 4 in Israel and the Middle East," *Scientific Reports* 10, no. 1 (2020): 1590.

77 **The job was labor intensive:** H. J. Su, S. C. Hwang, and W. H Ko, "Fusarial Wilt of Cavendish Bananas in Taiwan," *Plant Disease* 70 (1986): 814–18.

78 **other banana-growing nations:** Si-Jun Zheng et al., "New Geographical Insights of the Latest Expansion of *Fusarium oxysporum* f. sp. *cubense* Tropical Race 4 into the Greater Mekong Subregion," *Frontiers in Plant Science* 9 (2018): 457, https://doi.org/10 .3389/fpls.2018.00457.

78 **TR4 had arrived:** In 2021 the fungus was identified in Peru, prompting the nation to declare a national emergency. BananaLink, "Peru Declares National Emergency as TR4 Outbreak Is Confirmed," accessed July 21, 2022, https://www.bananalink.org .uk/news/peru-declares-national-emergency-as-tr4-outbreak-is-confirmed/.

78 **he is reluctant to claim:** Kema has been able to show such a relationship across TR4 incursions in the Mekong region in Southeast Asia. Kema, email communication with the author, August 2, 2022.

78 **hailed from Indonesia:** Maymon et al., "Origin and Current Situation."

78 **Kema is fairly certain:** Kema, interview.

78 **will be difficult:** Kema, email communication; AC van Westerhoven et al., "Uncontained spread of Fusarium wilt of banana threatens African food security," PLOS Pathogens 18 (2022): e1010769, https://doi.org/10.1371/journal.ppat.1010769.

78 **Forty thousand:** Luis Pocasangre, email communication with author, February 21, 2022; and Angelina Sanderson Bellamy, "Banana Production Systems: Identification of Alternative Systems for More Sustainable Production," *Ambio* 42, no. 3 (April 2013): 334–43.

78 **"Should TR4 invade":** Cahal Milmo, "'Noah's Ark' of the Fruit World Where the Banana Seeds of 1,600 Varieties Are Grown," *Independent*, April 7, 2014, accessed February 19, 2021, https://www.independent.co.uk/news/science/noah-s-ark-fruit -world-where-banana-seeds-1-600-varieties-are-grown-9239477.html.

79 **"In the developing world economies":** Sarah Gurr, interview by the author, January 26, 2022.

80 **genetic diversity:** Kema, interview.

80 **the fungus grows best:** At EARTH, Pocasangre says when there is a lot of rain, as there is in Limón, "it is impossible not to spray." "We use less," he says, about half of what a conventional grower in the region might use, which is how the fruit can be labeled sustainably grown rather than organic. Pocasangre, personal communication.

80 **if growers aren't spraying:** Rony Swennen, interview by the author, January 22, 2020.

80 **Yellow Sigatoka:** It was first identified in 1902 in Java, but then began to spread to Fiji, Australia, Ceylon, and other regions. Soluri, *Banana Cultures,* 104–7.

81 **copper residue:** Soluri, *Banana Cultures,* 108.

81 **While consumers:** Soluri, *Banana Cultures,* 215.

81 **"until it formed":** Steve Marquardt, "Pesticides, Parakeets, and Unions in the Costa Rican Banana Industry, 1938–1962," *Latin American Research Review* 37, no. 2 (2002): 3–36, quotes on 8, 11.

81 **"We spray workers":** Marquardt, "Pesticides, Parakeets," 3, 25, 28. For more about copper toxicity see Lamichhane et al., "Thirteen Decades of Antimicrobial Copper Compounds"; and Lori Ann Thrupp, "Long-Term Losses from Accumulation of Pesticide Residues: A Case of Persistent Copper Toxicity in Soils of Costa Rica," *Geoforum* 22, no. 1 (January 1, 1991): 1–15.

82 **toxic to humans and wildlife:** For more about fungicides and toxicity, see A. Chong Aguirre, "The Origin, Versatility and Distribution of Azole Fungicide Resistance in the Banana Black Sigatoka Pathogen *Pseudocercospora fijiensis,*" PhD diss., Wageningen University, Netherlands, 2016, https://doi.org/10.18174/387237; and William Henriques et al., "Agrochemical Use on Banana Plantations in Latin America: Perspectives on Ecological Risk," *Environmental Toxicology and Chemistry* 16, no. 1 (1997): 91–99.

82 **use of these fungicides:** Paul E. Verweij et al., "The One Health Problem of Azole Resistance in *Aspergillus fumigatus*: Current Insights and Future Research Agenda," *Fungal Biology Reviews* 34, no. 4 (December 2020): 202–14.

82 **remains a problem:** Madison Stewart, "The Deadly Side of America's Banana Obsession," Pulitzer Center, March 30, 2020, https://pulitzercenter.org/stories /deadly-side-americas-banana-obsession.

83 **"economic, political and social impacts":** Trevor Maynard, "Food System Shock: The Insurance Impacts of Acute Disruption to Global Food Supply, Emerging Risk Report 2015," Lloyd's, 2015, 1–30, https://legacy-assets.eenews.net/open_files/assets /2015/06/19/document_cw_02.pdf.

83 **with around 20 percent of:** Ravi P. Singh et al., "The Emergence of Ug99 Races of the Stem Rust Fungus Is a Threat to World Wheat Production," *Annual Review of Phytopathology* 49, no. 1 (September 8, 2011): 465–81.

83 **"a tiny part of a huge story":** Gurr, interview.

83 **a new host:** Helen N. Fones et al., "Threats to Global Food Security from Emerging Fungal and Oomycete Crop Pathogens," *Nature Food* 1, no. 6 (June 8, 2020): 332–42.

83 **"We can pretend":** Gurr, interview.

CHAPTER 5: NIGHT

85 **add up to several hundreds:** USGS, "What Do Bats Eat?" accessed December 7, 2021, https://www.usgs.gov/faqs/what-do-bats-eat?qt-news_science_products=0#qt -news_science_products.

86 **across the United States:** In less than ten years the fungus traveled across the Great Plains, killing and infecting bats from Manitoba to Washington State and as far south as Texas.

86 **90 percent of their population:** Tina L. Cheng et al., "The Scope and Severity of White-Nose Syndrome on Hibernating Bats in North America," *Conservation Biology* 35, no. 5 (2021): 1586–97; and Cheng et al., "Higher Fat Stores Contribute to Persistence of Little Brown Bat Populations with White-Nose Syndrome," *Journal of Animal Ecology* 88, no. 4 (April 1, 2019): 591–600.

88 **receive sounds between:** OSU Bio Museum, "Bat Sounds," Ohio State University, accessed December 7, 2021, https://u.osu.edu/biomuseum/2017/08/09/bat-sounds/.

88 **have adapted to this eating habit:** M. Lisandra Zepeda Mendoza et al., "Hologenomic Adaptations Underlying the Evolution of Sanguivory in the Common Vampire Bat," *Nature Ecology & Evolution* 2, no. 4 (2018): 659–68.

88 **flying fox:** The flying fox is native to the Philippines and is one of the many species of bats hunted for food and sold in markets as bush meat, a potential route for viral transfer from wildlife to humans. Pathogens endemic in bats may be precursors to viruses that cause pandemics in humans, including the recent SARS-CoV-2, though the route of the virus from bat to human, if it exists, has yet to be identified. Closer to home, bats may carry rabies, a rare, often fatal disease in humans if not treated in time.

89 **His work:** Christopher S. Richardson et al., "Thomas H. Kunz," *Physiological and Biochemical Zoology* 94, no. 4 (2021): 253–67; and Allen Kurta et al., "Obituary: Thomas Henry Kunz (1938–2020)," *Journal of Mammalogy* 101, no. 6 (2020): 1752–80.

89 **fastest mammals:** Gary F. McCracken et al., "Airplane Tracking Documents the Fastest Flight Speeds Recorded for Bats," *Royal Society Open Science* 3, no. 11 (December 7, 2021): 160398.

90 **set of caves:** Jonathan Reichard, interview by the author, July 19, 2019; and Elizabeth Kolbert, *The Sixth Extinction: An Unnatural History* (New York: Henry Holt, 2014).

91 **"The floor was littered":** Reichard, interview.

91 **Other scientists:** Riley Bernard, interview by the author, July 16, 2019; and Giorgia Auteri, "Are Bats Adapting to an Emergent Disease?" *Ecology and Evolution*, April 13, 2020, accessed July 30, 2020, https://ecoevocommunity.nature.com/posts /65734-are-bats-adapting-to-an-emergent-disease.

92 **colonized wings:** Vishnu Chaturvedi et al., "Morphological and Molecular Characterizations of Psychrophilic Fungus *Geomyces destructans* from New York Bats with White Nose Syndrome (WNS)," *PloS One* 5, no. 5 (May 2010): e10783.

92 **in and out of torpor:** Riley F. Bernard et al., "Identifying Research Needs to Inform White-Nose Syndrome Management Decisions," *Conservation Science and Practice* 2, no. 8 (August 30, 2020): e220.

93 **notorious for tolerating viruses:** Aaron T. Irving et al., "Lessons from the Host Defences of Bats, a Unique Viral Reservoir," *Nature* 589, no. 7842 (2021): 363–70; and Alice Latinne et al., "Coronaviruses in China," *Nature Communications* 11, no. 4235 (August 25, 2020).

93 **B-cells and T-cells:** Carol U. Meteyer, Daniel Barber, and Judith N. Mandl, "Pathology in Euthermic Bats with White Nose Syndrome Suggests a Natural Manifestation of Immune Reconstitution Inflammatory Syndrome," *Virulence* 3, no. 7 (November 15, 2012): 583–88.

93 **Marianne Moore:** Marianne Moore, interview by the author, January 20, 2021; and Marianne S. Moore et al., "Hibernating Little Brown Myotis (*Myotis lucifugus*) Show Variable Immunological Responses to White-Nose Syndrome," *PloS One* 8, no. 3 (2013): e58976.

93 **inflammatory response:** T. M. Lilley et al., "Immune Responses in Hibernating Little Brown Myotis (*Myotis lucifugus*) with White-Nose Syndrome," *Proceedings of the Royal Society B: Biological Sciences* 284, no. 1848 (February 8, 2017).

94 **immune system revives:** Moore, interview.

94 **rebound response:** Meteyer et al., "Pathology in Euthermic Bats," 3.

94 **immune reconstitution inflammation syndrome:** Meteyer et al., "Pathology in Euthermic Bats."

94 *then* **treated with the antiretrovirals:** Stuart M. Levitz, interview by the author, December 5, 2019.

95 **$23 billion:** Justin G. Boyles et al., "Economic Importance of Bats in Agriculture," *Science* 332, no. 6025 (2011): 41–42.

95 **bats are underappreciated contributors:** The threat of white nose syndrome is huge, but there is another looming threat—wind power. Wind turbines kill an estimated six hundred thousand to nine hundred thousand bats each year. The numbers will likely rise as the number and size of turbines increases. If white nose is an acute threat to hibernating North American bats, the turbines are potentially a chronic and global threat to migratory bats whose flight paths intersect with the turbines. See Daniel Y. Choi, Thomas W. Wittig, and Bryan M. Kluever, "An Evaluation of Bird and Bat Mortality at Wind Turbines in the Northeastern United States," *PloS One* 15, no. 8 (August 28, 2020): e0238034.

95 **roughly a thousandfold:** Andrew Cliff and Peter Haggett, "Time, Travel and Infection," *British Medical Bulletin* 69 (2004): 87–99.

95 **80 million international visitors:** US Travel Association, "US Travel and Tourism Overview," 2019, ustravel.org.

95 **traveler's clothing:** C. R. Wellings, R. A. McIntosh, and J. Walker, "*Puccinia striiformis* f. sp. *tritici* in Eastern Australia: Possible Means of Entry and Implications for Plant Quarantine," *Plant Pathology* 36, no. 3 (September 1987): 239–41.

96 **possibly Central Europe:** Jeffrey Foster, interview by the author, September 3, 2020; and Kevin Drees et al., "Phylogenetics of a Fungal Invasion: Origins and Widespread Dispersal of White-Nose Syndrome," 8 (March 2019): 1–15.

97 **None of:** Michael Campana et al., "White-Nose Syndrome Fungus in a 1918 Bat Specimen from France," *Emerging Infectious Disease Journal* 23, no. 9 (2017): 1611.

97 **background infection:** Marcus Fritze et al., "Determinants of Defence Strategies of a Hibernating European Bat Species towards the Fungal Pathogen *Pseudogymnoascus destructans*," *Developmental and Comparative Immunology* 119 (June 2021), https://doi.org/10.1016/j.dci.2021.104017.

97 **"the great divider":** Foster, interview.

98 **hopeful signs:** Kate E. Langwig et al., "Drivers of Variation in Species Impacts for a Multi-Host Fungal Disease of Bats," *Philosophical Transactions of the Royal Society B: Biological Sciences* 371, no. 1709 (December 5, 2016), https://doi.org/10.1098/rstb.2015.0456.

98 **significantly fatter:** Tina L. Cheng et al., "Higher Fat Stores Contribute to Persistence of Little Brown Bat Populations with White-Nose Syndrome," *Journal of Animal Ecology* 88, no. 4 (April 1, 2019): 591–600.

98 **less than 10 percent:** Cheng et al., "Scope and Severity of White-Nose Syndrome."

99 **In 1973:** Weiner, *The Beak of the Finch* (New York: Vintage, 1995), 43.

99 **Both Grants:** Emily Singer, "Watching Evolution Happen in Two Lifetimes," *Quanta Magazine*, September 22, 2016, https://www.quantamagazine.org/watching-evolution-happen-in-two-lifetimes-20160922/. For more see Joel Achenbach, "The People Who Saw Evolution," *Princeton Alumni Weekly*, April 23, 2014, accessed December 7, 2021, https://paw.princeton.edu/article/people-who-saw-evolution.

100 **they recalled:** Singer, "Watching Evolution Happen in Two Lifetimes."

100 **machinations of evolution:** Rosemary Grant and Peter R. Grant, "What Darwin's Finches Can Teach Us about the Evolutionary Origin and Regulation of Biodiversity," *BioScience* 53, no. 10 (October 2003): 965–75.

101 **Sixth Mass Extinction:** Gerardo Ceballos, Paul R. Ehrlich, and Peter H. Raven, "Vertebrates on the Brink as Indicators of Biological Annihilation and the Sixth Mass Extinction," *Proceedings of the National Academy of Sciences of the United States of America* 117, no. 24 (2020): 13596–602.

102 **a selective advantage:** Giorgia Auteri, interview by the author, February 26, 2020.

102 **"More bats than birds":** Auteri, interview.

104 **a few SNPs:** Giorgia G. Auteri and L. Lacey Knowles, "Decimated Little Brown Bats Show Potential for Adaptive Change," *Scientific Reports* 10, no. 1 (2020): 1–10.

104 **"None of these alleles":** Auteri, interview.

105 **evolve rapidly:** Craig L. Frank, April D. Davis, and Carl Herzog, "The Evolution of a Bat Population with White-Nose Syndrome (WNS) Reveals a Shift from an Epizootic to an Enzootic Phase," *Frontiers in Zoology* 16, no. 1 (2019): 1–9.

105 **"I don't think we are seeing":** Auteri, interview.

105 **prevailing idea:** The textbook example of this pathogen-host relationship is the release of the myxoma virus intended to kill European rabbits that had been released in Australia—to disastrous effect—in the late 1800s. Over a century later, evolution continues as the virus has evolved to suppress the immunity that has emerged in the rabbits. For more see Peter J. Kerr et al., "Evolutionary History and Attenuation of Myxoma Virus on Two Continents," *PLoS Pathogens* 8, no. 10 (October 4, 2012): e1002950.

105 **"If you walked down the stream":** Jamie Voyles, "Dr. Jamie Voyles: Epic Research Investigating Epidemics and Infectious Diseases in Wildlife," *People Behind the Science Podcast*, October 15, 2018, https://www.peoplebehindthescience.com/dr-jamie-voyles/.

106 **"were I a betting woman":** Jamie Voyles, interview by the author, March 31, 2020.

107 **frog and chytrid:** Vance Vredenburg, interview by the author, August 20, 2020.

107 **microbiomes are complicated:** Andrea J. Jani et al., "The Amphibian Microbiome Exhibits Poor Resilience Following Pathogen-Induced Disturbance," *ISME Journal* 15 (February 9, 2021): 1628–40; and Silas Ellison et al., "Reduced Skin Bacterial Diversity Correlates with Increased Pathogen Infection Intensity in an Endangered Amphibian Host," *Molecular Ecology* 28, no. 1 (2019): 127–40.

107 **"This really gives me hope":** Vance Vredenburg, interview by the author, December 18, 2020.

CHAPTER 6: RESISTANCE

113 **the odd survivor:** Gerald Barnes, unpublished memoir, received by author January 2021.

113 **billions of board feet:** Rocky Mountain Research Station, "Return of the King: Western White Pine Conservation and Restoration in a Changing Climate," accessed December 8, 2021, https://www.fs.usda.gov/rmrs/return-king-western-white-pine -conservation-and-restoration-changing-climate.

113 **tallest and broadest:** Sugar Pine Foundation, "Record Sugar Pines Discovered in the Sierra Nevada," accessed December 8, 2021, https://sugarpinefoundation.org /record-sugar-pines-discovered-in-the-sierra-nevada#comments.

114 **"whirlwind of destruction":** Donald Peattie, *A Natural History of North American Trees* (San Antonio, TX: Trinity Univesity Press, 2013), 46.

114 **in California's forests:** Louis T. Larsen and T. D. Woodbury, "Sugar Pine," USDA Bulletin no. 426 (Washington, DC, December 30, 1916), https://www.fs.fed.us/psw /publications/documents/usda_series/usda_bull426.pdf.

114 **from British Columbia:** Bohun B. Kinloch, "White Pine Blister Rust in North America: Past and Prognosis," *Phytopathology* 93 (March 7, 2003): 1044–47.

114 **a resistant parent:** Richard Bingham, "Blister Rust Resistant Western White Pine for the Inland Empire: The Story of the First 25 Years of the Research and Development Program," USDA, Forest Service, General Technical Report INT-146, June 1983.

114 **five-decade run:** Richard J. Klade, "Building a Research Legacy: The Intermountain Station 1911–1997," USDA, Forest Service, General Technical Report RMRS-GTR-184, 2006, https://doi.org/10.2737/RMRS-GTR-184.

115 **seedlings grown from:** Klade, "Building a Research Legacy."

115 **Gerald Barnes:** Barnes, unpublished memoir.

117 **fifty genes:** Brian P. McEvoy and Peter M. Visscher, "Genetics of Human Height," *Economics and Human Biology* 7, no. 3 (December 2009): 294–306.

117 **can be ephemeral:** H. M. Heybroek et al., "Resistance to Diseases and Pests in Forest Trees: Basic Biology and International Aspects of Rust Resistance in Forest Trees," in *Proceedings of the Third International Workshop on the Genetics of Host-Parasite Interactions in Forestry*, vol. 505 (Wageningen, Netherlands, 1980), 14–21.

117 **Bohun Kinloch:** Bohun Kinloch, "Sugar Pine: An American Wood," USDA (Washington, DC: US Government Printing Office, February 1984), https://books .google.com/books?id=5vaUmLYCgcoC&printsec=frontcover&source=gbs_ge _summary_r&cad=0#v=onepage&q&f=false.

118 **strains of rust:** J. N. King et al., "A Review of Genetic Approaches to the Management of Blister Rust in White Pines," *Forest Pathology* 40, no. 3–4 (2010): 292–313; and Richard A. Sniezko, Jeremy S. Johnson, and Douglas P. Savin, "Assessing the Durability, Stability, and Usability of Genetic Resistance to a Non-native Fungal Pathogen in Two Pine Species," *Plants, People, Planet* 2, no. 1 (2020): 57–68.

118 **major genes *and* multiple genes:** Richard A. Sniezko and Jennifer Koch, "Breeding Trees Resistant to Insects and Diseases: Putting Theory into Application," *Biological Invasions* 19, no. 11 (November 20, 2017): 3377–400.

119 **must be representative:** Richard Sniezko, interview by the author, April 2020.

120 **Collectors seek seeds:** Haley Smith, interview by the author, February 17, 2022.

Smith is the seed program coordinator and horticulturist at the Dorena Genetic Resource Center.

122 **"really really good":** Sniezko, interview.

123 **Whitebark restoration:** Robert E. Keane et al., *A Range-Wide Restoration Strategy for Whitebark Pine (*Pinus albicaulis*)*, Gen. Tech. Rep. RMRS-GTR-279 (Fort Collins, CO: US Department of Agriculture, Forest Service, Rocky Mountain Research Station, 2012).

123 **fire can also kill:** Bob Keane, A. D. Bower, and Sharon Hood, "A Burning Paradox: Whitebark Is Easy to Kill but Also Dependent on Fire," *Nutcracker Notes* 38 (2020): 7–8, 34.

123 **Soil type and soil microbes:** Keane et al., *Range-Wide Restoration Strategy*; Cathy L. Cripps et al., "Inoculation and Successful Colonization of Whitebark Pine Seedlings with Native Mycorrhizal Fungi under Greenhouse Conditions," in *The Future of High-Elevation, Five-Needle White Pines in Western North America*, Proceedings of the High-Five Symposium, Missoula, MT, June 28–30, 2010, ed. Robert E. Keane, Diana F. Tomback, Michael P. Murray, and Cyndi M. Smith (Fort Collins, CO: USDA, Forest Service, Rocky Mountain Research Station, 2011).

123 **soil inoculums:** Cathy L. Cripps and Eva Grimme, "The Future of High-Elevation, Five-Needle White Pines in Western North America: Proceedings of the High Five Symposium," Keane et al., eds., in *Future of High Elevation White Pines*.

124 **hundreds of millions of trees:** Jad Daley, "Save Our Summits," American Forests, December 20, 2020, accessed December 8, 2021, https://americanforests.medium .com/save-our-summits-bc1721cee95a.

124 **a large consortium:** Kristian A. Stevens et al., "Sequence of the Sugar Pine Megagenome," *Genetics* 204, no. 4 (December 1, 2016): 1613–26.

125 **By length and weight:** Allison Piovesan et al., "On the Length, Weight and GC Content of the Human Genome," *BMC Research Notes* 12, no. 1 (February 27, 2019): 106.

125 **thirty-one billion base-pairs:** Kristian A. Stevens et al., "Sequence of the Sugar Pine Megagenome," *Genetics* 204, no. 4 (December 1, 2016): 1613–26.

126 **Sequencing the genome:** David Neale, interview by the author, January 12, 2021.

126 **human genome cost:** National Human Genome Research Institute, "The Cost of Sequencing a Human Genome," accessed December 8, 2021, https://www.genome .gov/about-genomics/fact-sheets/Sequencing-Human-Genome-cost.

127 **the ultimate success:** Tomback, interview by the author, May 13, 2022.

CHAPTER 7: DIVERSITY

128 **called landraces:** Julian Ramirez-Villegas et al., "State of Ex Situ Conservation of Landrace Groups of 25 Major Crops," *Nature Plants* 8 (2022): 491–99. For a more detailed discussion of landraces see Francesc Casañas et al., "Toward an Evolved Concept of Landrace," *Frontiers in Plant Science* 8 (February 8, 2017): 1–7.

129 **genes dropped:** Adi B. Damania, "History, Achievements, and Current Status of Genetic Resources Conservation," *Agronomy Journal* 100, no. 1 (January 1, 2008): 9–21.

129 **between 1903 and 1983:** Cary Fowler, *Seeds on Ice: Svalbard and the Global Seed Vault* (Westport, CT: Prospecta, 2016), 82. See also, about seed banking, Marci Baranski,

"Seed Banking 1979–1994," Embryo Project Encyclopedia, January 28, 2014, https://embryo.asu.edu/pages/seed-banking-1979-1994.

130 **"What my team and I found":** Cary Fowler, "Seeds on Ice," *American Scientist* 104, no. 5 (2016): 304.

130 **Global Seed Vault:** Fowler, *Seeds on Ice.*

130 **Nikolai Vavilov:** Gary Nabhan, "How Nikolay Vavilov, the Seed Collector Who Tried to End Famine, Died of Starvation," Splendid Table, accessed February 25, 2021, https://www.splendidtable.org/story/2013/10/17/how-nikolay-vavilov-the-seed-collector-who-tried-to-end-famine-died-of-starvation.

131 **cause of death:** Marci Baranski, "Nikolai Ivanovic Vavilov (1887–1943)," Embryo Project Encyclopedia, March 15, 2014, https://embryo.asu.edu/pages/nikolai-ivanovic-vavilov-1887-1943.

131 **seven million:** Sara Peres, "Saving the Gene Pool for the Future: Seed Banks as Archives," *Studies in History and Philosophy of Science Part C: Studies in History and Philosophy of Biological and Biomedical Sciences* 55 (2016): 96–104.

131 **Laboratory for Genetic Resources Preservation:** R. J. Griesbach, "150 Years of Research at the United States Department of Agriculture: Plant Introduction and Breeding," USDA, Agricultural Research Service, June 2013, https://www.ars.usda.gov/ARSUserFiles/oc/np/150YearsofResearchatUSDA/150YearsofResearchatUSDA.pdf.

131 **In the 1990s:** Griesbach, "150 Years." USDA seed searches can be done at this site: https://www.grin-global.org/.

131 **seeds and germplasm:** Simran Sethi, "This Colorado Vault Is Keeping Your Favorite Foods from Going Extinct," Counter, March 5, 2018, https://thecounter.org/seed-banks-biodiversity-preservation/. There is germplasm and seed held elsewhere around the country. There are fruits and nuts in Corvallis, Oregon. A site in Geneva, New York, affiliated with Cornell University preserves apple, cherry, and grape diversity. The university is also home to one of the newer collections, a hemp seed repository. Seeds and plants are distributed to researchers and breeders around the world who seek to develop a crop that can withstand drought, disease, and pests.

132 **"the thin green line":** Kaine Korzekwa, "The Necessity of Finding, Conserving Crop Wild Relatives," Science Daily, accessed April 25, 2022, https://www.sciencedaily.com/releases/2018/09/180926082718.htm; and Eric Debner, "Ames Seed Bank Saves for Future," *Iowa State Daily*, October 2, 2012.

132 **"Many genebanks":** Fowler, *Seeds on Ice,* 48.

132 **millions of seeds:** Fowler, "Seeds on Ice," 304.

132 **most susceptible states:** American Phytopathological Society, accessed December 14, 2021, https://www.apsnet.org/members/leadership/history/Documents/1908-1918_right.pdf.

133 **he witnessed:** Charles Mann, *The Wizard and the Prophet: Two Remarkable Scientists and Their Dueling Visions to Shape Tomorrow's World* (New York: Vintage, 2018), 108.

133 **a plant pathologist:** Borlaug did his graduate studies at the University of Minnesota, where he studied plant pathology under Elvin Stakman. During his graduate years Borlaug worked on fungal pathogens other than rust. But years later, when Stakman was asked to work on the problem of stem rust in Mexico, he recruited Borlaug—who at the time was working for DuPont de Nemours. For more see Mann, *Wizard and Prophet*; and Richard Zeyen et al., "Norman Borlaug: Plant Pathologist/Humanitarian," *APSnet Feature Articles*, 2000, https://doi.org/10.1094/apsfeature-2009-12.

133 **"an exercise"**: Mann, *Wizard and Prophet*, 128.

133 **Borlaug captured:** In awarding Borlaug the Nobel Peace Prize, the Nobel committee chairwoman, Aase Lionaes, said of him, "More than any other single person of this age, he has helped to provide bread for a hungry world. We have made this choice in the hope that providing bread will also give the world peace." Lionaes, "The Nobel Peace Prize 1970 Presentation Speech," NobelPrize.org, accessed December 14, 2021, https://www.nobelprize.org/prizes/peace/1970/ceremony-speech/.

133 **Borlaug's wheat:** Borlaug Global Rust Initiative, accessed December 14, 2021, https://bgri.cornell.edu/.

134 **"distribution of":** Norman Borlaug, "Foreword," in "Sounding the Alarm on Global Stem Rust," Expert Panel on the Stem Rust Outbreak in Eastern Africa, May 29, 2005, https://bgri.cornell.edu/wp-content/uploads/2020/12/Sounding AlarmGlobalRust.pdf.

134 **"Plant disease immunity":** Sarah Gurr, interview by the author, January 26, 2022.

135 **breeders have sought:** Pablo D. Olivera, Matthew N. Rouse, and Yue Jin, "Identification of New Sources of Resistance to Wheat Stem Rust in *Aegilops* spp. in the Tertiary Genepool of Wheat," *Frontiers in Plant Science* 9 (November 22, 2018): 1719; Dag Terje Filip Endresen et al., "Sources of Resistance to Stem Rust (Ug99) in Bread Wheat and Durum Wheat Identified Using Focused Identification of Germplasm Strategy," *Crop Science* 52, no. 2 (2012): 764–73; and CGIAR Genebank Platform, "Wheat ('Triticum' spp.) Is the World's Most Important Food Crop," accessed February 25, 2021, https://www.genebanks.org/resources/crops/wheat/.

135 **"Every time you breed":** Gurr, interview.

136 **such engineered wheat crops:** Guotai Yu et al., "*Aegilops sharonensis* Genome-Assisted Identification of Stem Rust Resistance Gene Sr62," *Nature Communications* 13, no. 1607 (March 25, 2022). Flour from wheat engineered to resist drought was approved in Argentina in 2020. So far it is the first and only engineered wheat approved in the world. In 2013 engineered wheat was found growing in a field in Oregon. Dan Charles, "GMO Wheat Found in Oregon Field: How Did It Get There?" *The Salt*, NPR, May 30, 2013, https://tinyurl.com/bde68zhj.

136 **"our number one weapon":** Gurr, interview

136 **enzyme necessary for fungi:** Zeina A. Kanafani and John R. Perfect, "Resistance to Antifungal Agents: Mechanisms and Clinical Impact," *Clinical Infectious Diseases* 46 (2008): 120–28.

136 **fatality rates:** Jan W. M. Van Der Linden et al., "Clinical Implications of Azole Resistance in *Aspergillus fumigatus*," 17, no. 10 (2012): 2007–9; and P. P. A. Lestrade et al., "Triazole Resistance in *Aspergillus fumigatus*: Recent Insights and Challenges for Patient Management," *Clinical Microbiology and Infection* 25, no. 7 (2019): 799–806.

136 **In 2007:** Paul Verweij et al., "Triazole Fungicides and the Selection of Resistance to Medical Triazoles in the Opportunistic Mould *Aspergillus fumigatus*," *Pest Management Science* 69, no. 2 (2013): 165–70; and Paul E. Verweij, Emilia Mellado, and Willem J. G. Melchers, "Multiple-Triazole–Resistant Aspergillosis," *New England Journal of Medicine* 356, no. 14 (April 5, 2007): 1481–83.

137 **azoles in the environment:** Verweij, Mellado, and Melchers, "Multiple-Triazole–Resistant Aspergillosis"; and Paul E. Verweij et al., "Azole Resistance in *Aspergillus fumigatus*: A Side-Effect of Environmental Fungicide Use?" *Lancet Infectious Diseases* 9, no. 12 (2009): 789–95.

137 **on tulip bulbs:** Katie Dunne et al., "Intercountry Transfer of Triazole-Resistant

Aspergillus fumigatus on Plant Bulbs," *Clinical Infectious Diseases* 65, no. 1 (March 29, 2017): 147–56; and Daisuke Hagiwara, "Isolation of Azole-Resistant *Aspergillus fumigatus* from Imported Plant Bulbs in Japan and the Effect of Fungicide Treatment," *Journal of Pesticide Science* 45, no. 3 (August 20, 2020): 147–50.

137 **To protect bulbs:** Paul E. Verweij et al., "The One Health Problem of Azole Resistance in *Aspergillus fumigatus*: Current Insights and Future Research Agenda," *Fungal Biology Reviews* 34, no. 4 (December 2020): 202–14. For more about tulips, azoles, and aspergillus see Maryn McKenna, "When Tulips Kill," *Atlantic*, November 15, 2018.

137 **similar mutations:** Paul E. Verweij et al., "One Health Problem"; Caroline Burks et al., "Azole-Resistant *Aspergillus fumigatus* in the Environment: Identifying Key Reservoirs and Hotspots of Antifungal Resistance," *PLoS Pathogens* 17, no. 7: e1009711 (July 29, 2021); and Johanna Rhodes et al., "Population Genomics Confirms Acquisition of Drug-Resistant *Aspergillus fumigatus* Infection by Humans from the Environment," *Nature Microbiology* 7 (April 25, 2022): 663–74.

137 **In the United States:** Toda Mitsuru et al., "Trends in Agricultural Triazole Fungicide Use in the United States, 1992–2016, and Possible Implications for Antifungal-Resistant Fungi in Human Disease," *Environmental Health Perspectives* 129, no. 5 (April 26, 2022): 55001. For updated information see Centers for Disease Control and Prevention, "Antifungal-Resistant Aspergillus," accessed April 28, 2022, https://www.cdc.gov/fungal/diseases/aspergillosis/antifungal-resistant.html.

137 **target multiple sites:** Gero Steinberg and Sarah J. Gurr, "Fungi, Fungicide Discovery and Global Food Security," *Fungal Genetics and Biology* 144 (2020): 103476.

138 **Scientists working:** Gert Kema, interview by the author, September 15, 2020; and Rony Swennen, interview by the author, January 22, 2020.

139 **in the ITC:** The ITC collection is stored at Katholieke Universiteit Leuven (KU Leuven). International Musa Germplasm Transit Centre, accessed July, 25, 2022, https://www.bioversityinternational.org/banana-genebank/.

139 **ITC sends:** Ines Van den Houwe et al., "Safeguarding and Using Global Banana Diversity: A Holistic Approach," *CABI Agriculture and Bioscience* 1, no. 1 (December 22, 2020): 15.

140 **virtually seedless:** ProMusa, "Domestication of the Banana," accessed December 14, 2021, https://www.promusa.org/Domestication+of+the+banana.

140 **"You *can* breed":** Swennen, interview.

140 **tedious process:** Allan Brown et al., "Bananas and Plantains (*Musa* spp.)," in *Genetic Improvement of Tropical Crops*, ed. H. Campos and P. D. S. Caligari (Springer International, 2017), 227.

140 **"We have tens of thousands":** Swennen, interview.

140 **Swennen describes:** "3rd PlantB+B Online Café_Rony Swennen," YouTube, October 15, 2021, accessed November 12, 2021, https://www.youtube.com/watch?v=yAfAMBUxCy0.

CHAPTER 8: RESURRECTION

143 **chestnut wood that had grown:** University of Massachusetts, Amherst, "Faculty Revive Tradition by Marking Centennial of Metawampe Hike on Mt. Toby," News and Media Relations, October 16, 2007, https://www.umass.edu/archivenewsoffice/article/faculty-revive-tradition-marking-centennial-metawampe-hike-mt-toby.

143 **"clear the forest":** Jesse Caputo and Tony D'Amato, "Mount Toby Demonstration For-

est Management Plan," Spring 2006, University of Massachusetts, Amherst, https://eco
.umass.edu/wp-content/uploads/file/pdfs/Mount_Toby_Final_Plan_May_24.pdf.

143 **All true chestnuts:** Taylor Perkins, email communication with author, November
17, 2022; Ping Lang et al., "Molecular evidence for an Asian origin and a unique
westward migration of species in the genus Castanea via Europe to North America," *Molecular Phylogenetics and Evolution* 43 (2007): 49–59; BF Zhou et al., "Phy-
logenomic analyses highlight innovation and introgression in the continental
radiations of Fagaceae across the Northern Hemisphere," *Nature Communication* 13
(2022): 1320, https://doi.org/10.1038/s41467-022-28917-1.

144 **"not a single tree":** D. Fairchild, "The Discovery of the Chestnut Bark Disease
in China," *Science* 38, no. 974 (August 1913): 297–99. For more about Frank Meyer
see Frank N. Meyer, "Archives III FNM," 2012, 1906–14, Archives of the Arnold
Arboretum, Harvard University, Cambridge, MA, https://arboretum.harvard.edu
/wp-content/uploads/2020/07/III_FNM_2012.pdf.

144 **Graves, a curator:** Susan Freinkel, *American Chestnut: The Life, Death, and Rebirth of a
Perfect Tree* (Berkeley: University of California Press, 2007), 96.

145 **he cut catkins:** For details on early breeding efforts see Freinkel, *American Chestnut*;
and Richard A. Jaynes and Arthur Graves, "Connecticut Hybrid Chestnuts and Their
Culture" (New Haven, CT: Connecticut Agricultural Experiment Station, 1963).

145 **thirty years:** Henry Svenson, "Arthur Harmount Graves," *Bulletin of the Torrey Botanical Club* 90, no. 5 (September–October 1963): 332–36; and Jaynes and Graves, "Connecticut Hybrid Chestnuts."

146 **None were:** Freinkel, *American Chestnut*, 100.

146 **more difficult task:** R. A. Jaynes, "Selecting and Breeding Blight Resistant Chestnut
Trees," in *Proceedings of the American Chestnut Symposium*, ed. William L Macdonald et
al. (Morgantown: West Virginia University, 1978), 4–6, https://www.fs.fed.us/nrs/
pubs/jrnl/1978/ne_1978_macdonald_chestnutproc.pdf.

146 **Five years later:** Sandra L. Anagnostakis, "Chestnut Breeding in the United States
for Disease Insect Resistance," *Plant Disease* 96, no. 10 (October 2012): 1392–403; and
Jaynes, "Selecting and Breeding Blight Resistant Chestnut Trees," 5.

147 **Theoretically by the third backcross:** Charles R. Burnham, "The Restoration of
the American Chestnut," *American Scientist* 76 (1988): 478–87.

148 **Burnham knew:** Freinkel, *American Chestnut*, 134.

150 **more than three genes:** Jared W. Westbrook et al., "Optimizing Genomic Selection
for Blight Resistance in American Chestnut Backcross Populations: A Trade-off with
American Chestnut Ancestry Implies Resistance Is Polygenic," *Evolutionary Applications* 13, no. 1 (January 29, 2020): 31–47.

150 **thirty-one scientists:** Margaret Staton et al., "The Chinese Chestnut Genome:
A Reference for Species Restoration," *BioRxiv Preprint*, April 22, 2019, https://doi
.org/10.1101/615047.

150 **Jared Westbrook:** Jared Westbrook, personal communication, February 14, 2022.

151 **the BC3F3s:** Westbrook et al., "Optimizing Genomic Selection."

151 **hypovirulent:** Freinkel, *American Chestnut*, 111–28.

152 **in orchards:** D. L. Nuss, "Biological Control of Chestnut Blight: An Example of
Virus-Mediated Attenuation of Fungal Pathogenesis," *Microbiological Reviews* 56, no.
4 (December 1992): 561–76; and Ursula Heiniger and Daniel Rigling, "Biological
Control of Chestnut Blight in Europe," *Annual Review of Phytopathology* 32, no. 1 (September 1, 1994): 581–99.

152 **It worked:** N. K. Van Alfen et al., "Chestnut Blight: Biological Control by Transmissible Hypovirulence in *Endothia parasitica*," *Science* 189, no. 4206 (September 12, 1975): 890–91.

152 **different strains:** Nuss, "Biological Control of Chestnut Blight," 563.

152 **didn't release oxalate:** Evelyn A. Havir and Sandra L. Anagnostakis, "Oxalate Production by Virulent but Not by Hypovirulent Strains of *Endothia parasitica*," *Physiological Plant Pathology* 23, no. 3 (1983): 369–76.

153 **hinder common fungi:** SUNY College of Environmental Science and Forestry, "Restoring the American Chestnut: The Search for Blight Resistant-Enhancing Genes," Syracuse, NY, accessed April 26, 2022, https://www.esf.edu/chestnut/genes.htm.

153 **In his book:** Daniel Charles, *Lords of the Harvest: Biotech, Big Money, and the Future of Food* (Cambridge, MA: Perseus, 2001), 24.

154 **sixteen million hectares:** George Silva, "Global Genetically Modified Crop Acres Increase amid Concerns," Michigan State University Extension, East Lansing, December 12, 2017, https://www.canr.msu.edu/news/global_genetically_modified _crop_acres_increase_amid_concerns.

154 **banned or restricted:** Genetic Literacy Project, "Where Are GMO Crops and Animals Approved and Banned?" accessed December 9, 2021, https://geneticliteracy project.org/gmo-faq/where-are-gmo-crops-and-animals-approved-and-banned/.

154 **in-depth review:** Danny Hakim, "Doubts about the Promised Bounty of Genetically Modified Crops," *New York Times*, October 29, 2016.

154 **crops engineered:** Edward D. Perry et al., "Genetically Engineered Crops and Pesticide Use in U.S. Maize and Soybeans," *Science Advances* 2, no. 8 (August 31, 2016): e1600850; and Akhter U. Ahmed et al., "The Impacts of GM Foods: Results from a Randomized Controlled Trial of Bt Eggplant in Bangladesh," *American Journal of Agricultural Economics* 103, no. 4 (November 13, 2020): 1186–1206.

155 **first blight-tolerant American chestnut:** Bo Zhang et al., "A Threshold Level of Oxalate Oxidase Transgene Expression Reduces *Cryphonectria parasitica*-Induced Necrosis in a Transgenic American Chestnut (*Castanea dentata*) Leaf Bioassay," *Transgenic Research* 22, no. 5 (2013): 973–82; and Andrew E. Newhouse et al., "Transgenic American Chestnuts Show Enhanced Blight Resistance and Transmit the Trait to T1 Progeny," *Plant Science* 228 (November 1, 2014): 88–97.

156 **three-hundred-page petition:** See a summary of the petition here: William A. Powell et al., "Petition for Determination of Nonregulated Status for Blight-Tolerant Darling 58 American Chestnut," SUNY College of Environmental Science and Forestry, Syracuse, n.d., https://www.esf.edu/chestnut/documents/petition_executive _summary.pdf.

156 **"development through":** "Comment from Global Forest Coalition," Regulations. gov, accessed December 14, 2021, https://www.regulations.gov/comment/APHIS -2020-0030-4498. For more about the Sierra Club's position see Kate Morgan, "The Demise and Potential Revival of the American Chestnut," *Sierra*, February 25, 2021, https://www.sierraclub.org/sierra/2021-2-march-april/feature/demise-and-potential -revival-american-chestnut.

156 **a model:** Here is what Jared Westbrook has to say: "I think the greatest hope for chestnut is in the combination of traditional backcross breeding to bring in some genes from Chinese chestnut to enhance resistance, while also breeding some of our most resistant trees with the transgenic Darling 58 lines to create 'stacked resistance' lines. Adding in CRISPR approaches may offer additional hope. However, we do not yet

have a detailed enough knowledge of how resistance works in Asian chestnut species to know which specific edits to make." Jared Westbrook, email communication with author, February 14, 2021.

157 **crop plants:** Syed Shan-e-ali Zaidi et al., "Engineering Crops of the Future: CRISPR Approaches to Develop Climate-Resilient and Disease-Resistant Plants," *Genome Biology* 21 (November 30, 2020): 289. These crops are in various stages of development.

157 **modify bananas:** Leena Tripathi, Valentine O. Ntui, and Jaindra N. Tripathi, "CRISPR/Cas9-Based Genome Editing of Banana for Disease Resistance," *Current Opinion in Plant Biology* 56, no. 1 (2020): 118–26; and Jaindra N. Tripathi et al., "CRISPR/Cas9 Editing of Endogenous Banana Streak Virus in the B Genome of Musa spp. Overcomes a Major Challenge in Banana Breeding," *Communications Biology* 2, 46 (2019). For more about CRISPR in crops see Zaidi et al., "Engineering Crops of the Future." The use of Cas9 may or may not require the insertion of foreign DNA depending on the process, although in some applications, if used, it can be bred out of the final product, leaving a plant that has been modified yet lacks foreign genes. Mollie Rappe, "CRISPR Plants: New Non-GMO Method to Edit Plants," CALS News, North Carolina State University, Raleigh, May 11, 2020, accessed December 14, 2021, https://cals.ncsu.edu/news/crispr-plants-new-non-gmo-method-to-edit-plants/; and Janina Metje-Sprink et al., "DNA-Free Genome Editing: Past, Present and Future," *Frontiers in Plant Science* 9 (January 14, 2019): 1957.

157 **resistance gene:** James Dale et al., "Transgenic Cavendish Bananas with Resistance to Fusarium Wilt Tropical Race 4," *Nature Communications* 8, no. 1 (2017): 1496.

158 **"the lines are still classified":** James Dale, email communication with the author, November 12, 2021.

158 **in Cavendish bananas:** Amy Maxmen, "CRISPR Could Save Bananas from Fungus," *Nature* 574 (October 3, 2021), https://media.nature.com/original/magazine-assets/d41586-019-02770-7/d41586-019-02770-7.pdf; and James Dale, email communication, November, 17, 2021.

158 **not be regulated:** Tripathi, Ntui, and Tripathi, "CRISPR/Cas9-Based Genome Editing."

158 **how the fruit will be received:** Several strategies likely in combination will be needed. One, which Swennen and others including Pocasangre favor, involves biological controls, that is, adding or cultivating bacteria and fungi that provide nutrients or protective chemicals or that can provoke a plant's natural defenses. Naturally occurring soil microbes can provide a treasure trove of enzymes that can make nutrients more available to plants and may help fend off opportunistic pathogens—much like our gut microbes. See also Manoj Kaushal, Rony Swennen, and George Mahuku, "Unlocking the Microbiome Communities of Banana (*Musa* spp.) under Disease Stressed (Fusarium Wilt) and Non-Stressed Conditions," *Microorganisms* 8, no. 3 (2020): 443. For further reading see David Montgomery and Anne Biklé, *The Hidden Half of Nature* (New York: W. W. Norton, 2016).

158 **it would be centuries:** Jared Westbrook, email communication with author, April 26, 2020. Westbrook credits Sara Fitzsimmons with the calculation; for more see Sara F. Fitzsimmons, "Magnitude of American Chestnut Restoration and the Roles of TAFC Chapters over the Next 40+ Years," YouTube, November 9, 2021, https://www.youtube.com/watch?v=PY6Ua1haBao.

159 **let the forest be:** Tom Wessels, interview by the author, June 9, 2020.

CHAPTER 9: CERTIFICATION

160 **artist James Mahoney:** James Mahoney, "Views of the Famine," *Illustrated London News*, February 13, 1847, accessed August 9, 2016, http://viewsofthefamine.wordpress .com/illustrated-london-news/sketches-in-the-west-of-ireland/.

160 **it traveled:** The disease likely traveled from South America to the United States and then to Europe. Amanda C. Saville and Jean B. Ristaino, "Global Historic Pandemics Caused by the FAM-1 Genotype of *Phytophthora infestans* on Six Continents," *Scientific Reports* 11 (June 11, 2021), https://doi.org/10.1038/s41598-021-90937-6.

161 **geographic barriers:** Bruce S. Lieberman, "The Geography of Evolution and the Evolution of Geography," *Evolution: Education and Outreach* 5, no. 4 (2012): 521–25.

161 **"life's richness":** Elizabeth Kolbert, *The Sixth Extinction: An Unnatural History* (New York: Henry Holt, 2014), 195–98.

161 **turned to rot:** Jean Beagle, Jean Ristaino, and Donald H. Pfister, "'What a Painfully Interesting Subject': Charles Darwin's Studies of Potato Late Blight," *BioScience* 66, no. 12 (2016): 1035.

162 **"I fear this decides":** Beagle, Ristano, and Pfister, "What a Painfully Interesting Subject."

162 **other wild potatoes:** Hari S. Karki, Shelly H. Jansky, and Dennis A. Halterman, "Screening of Wild Potatoes Identifies New Sources of Late Blight Resistance," *Plant Disease* 105, no. 2 (August 5, 2020): 368–76. Recently, origins of *P. infestans* have been traced from South America to the eastern United States and then to Europe. From there it spread to Africa, India, China, and Australia most likely with British colonialists. For more see Saville and Ristaino, "Global Historic Pandemics."

162 **colleague of Darwin's:** Beagle, Ristaino, and Pfister, in "What a Painfully Interesting Subject," suggest that the fungus was later confirmed as the cause by German microbiologist and mycologist Anton de Bary, who is also considered a founding scientist of plant pathology and modern mycology. See also U. Kutschera and U. Hossfeld, "Physiological Phytopathology: Origin and Evolution of a Scientific Discipline," *Journal of Applied Botany and Food Quality* 85 (2012): 1–5.

163 **has exploded:** Andrew M. Liebhold et al., "Live Plant Imports: The Major Pathway for Forest Insect and Pathogen Invasions of the US," *Frontiers in Ecology and the Environment* 10, no. 3 (February 18, 2012): 135–43; and Faith Campbell, "Living Plant Imports: Scientists Try to Counter Longstanding Problems," Center for Invasive Species Prevention, December 21, 2021, accessed February 18, 2022, https://www .nivemnic.us/?p=2935.

164 **seventeen such stations:** USDA, Economic Research Service, "Agricultural Trade," last updated August 20, 2019, accessed February 11, 2021, https://www.ers.usda.gov /data-products/ag-and-food-statistics-charting-the-essentials/agricultural-trade/.

165 **inspection teams identify:** Deborah G. McCullough et al., "Interceptions of Nonindigenous Plant Pests at US Ports of Entry and Border Crossings over a 17-Year Period," *Biological Invasions* 8 (January 20, 2006), https://doi.org/10.1007/s10530-005 -1798-4.

165 **fire salamander:** A. Martel et al., "Recent Introduction of a Chytrid Fungus Endangers Western Palearctic Salamanders," *Science* 346, no. 6209 (2014): 630–31; Simon J. O'Hanlon et al., "Recent Asian Origin of Chytrid Fungi Causing Global Amphibian Declines," *Science* 360, no. 6389 (May 11, 2018): 621–27; and An Martel et al., "*Batrachochytrium salamandrivorans* sp. nov. Causes Lethal Chytridiomycosis in Amphibi-

ans," *Proceedings of the National Academy of Sciences of the United States of America* 110, no. 38 (September 3, 2013): 15325–29.

166 **Asian newts:** Martel et al., "*Batrachochytrium salamandrivorans* sp. nov."

166 **in Vietnam:** Matthew C. Fisher and Trenton W. J. Garner, "Chytrid Fungi and Global Amphibian Declines," *Nature Reviews Microbiology* 18, no. 6 (2020): 332–43.

166 **"ancient barriers":** Martel et al., "*Batrachochytrium salamandrivorans* sp. nov."

166 **a hotspot:** Tiffany A. Yap et al., "Averting a North American Biodiversity Crisis: A Newly Described Pathogen Poses a Major Threat to Salamanders via Trade," *Science* 349, no. 6247 (2015): 481–82; and "AmphibiaWeb," accessed February 16, 2022, https://amphibiaweb.org/. Collecting and keeping some indigenous species of salamanders is illegal in some states.

166 **many of them from Asia:** Martel et al., "Recent Introduction of a Chytrid Fungus"; and Defenders of Wildlife, "SOS—Save Our Salamanders," July 28, 2015, accessed December 12, 2021, https://defenders.org/blog/2015/07/sos-save-our-salamanders.

167 **"morally irresponsible":** C. Gascon et al., eds., *Amphibian Conservation Action Plan* (Gland, Switzerland: IUCN/SSC Amphibian Specialist Group, 2007), 4.

167 **She was often asked:** Margaret Krebs et al., "Narrative: Stopping a Disease from Becoming a Crisis," National Socio-Environmental Synthesis Center, May 16, 2017, https://www.sesync.org/resources/stopping-wildlife-disease-becoming-crisis -collaborative-leadership-success-story; and Karen Lips, interview by the author, March 13, 2020.

168 **Rabb helped:** Joseph C. Mitchell, Joseph R. Mendelson, and Margaret M. Stewart, "George Bernard Rabb," *Copeia* 105, no. 3 (2017): 592–98.

168 **"ensure traded amphibians":** US Fish and Wildlife, "Injurious Wildlife Species; Review of Information Concerning a Petition to List All Live Amphibians in Trade as Injurious unless Free of *Batrachochytrium dendrobatidis*," *Federal Register*, September 17, 2010, accessed December 16, 2021, https://www.federalregister .gov/documents/2010/09/17/2010-23039/injurious-wildlife-species-review-of -information-concerning-a-petition-to-list-all-live-amphibians.

168 **In February 2001:** For more about the outbreak see CNN, "Foot-and-Mouth Crisis Timetable," CNN.com/World, August 7, 2001, http://edition.cnn.com/2001/WORLD /europe/UK/04/11/fandm.timeline/; and Alan Colwell, "Foot-and-Mouth Disease Keeps Hikers Indoors," *New York Times*, March 18, 2001.

169 **Britain and in the Netherlands:** Anthony Browne, "Protesters March to Halt Mass Slaughter," *Guardian*, April 21, 2001.

169 **£8 billion:** Steve Malakowsky, "Billions of Reasons (Dollars) to Keep Foot-and-Mouth at Bay," National Hog Farmer, April 19, 2017, accessed December 16, 2021, https:// www.nationalhogfarmer.com/animal-health/billions-reasons-dollars-keep-foot-and -mouth-bay.

169 **In 1929:** Alejandro Segarra and Jean Rawson, "CRS Report for Congress," 2001, https://28xeuf2otxva18q7lx1uemec-wpengine.netdna-ssl.com/https://28xeuf2ot xva18q7lx1uemec-wpengine.netdna-ssl.com/wp-content/uploads/assets/crs/RS 20890.pdf.

169 **"We're one cow away":** CNN, "Foot-and-Mouth Disease Precautions," CNN.com /World, March 14, 2001, https://www.cnn.com/2001/WORLD/europe/03/14/foot .mouth.measures/.

169 **President George W. Bush:** CNN, "Foot-and-Mouth Disease Precautions"; and Segarra and Rawson, "CRS Report for Congress."

170 **"response exercise":** Department for Environment, Food, and Rural Affairs,

United Kingdom, "National Foot and Mouth Disease Exercise Evaluation and Lessons Identified Report," October 9, 2018, https://assets.publishing.service.gov .uk/government/uploads/system/uploads/attachment_data/file/925804/foot-and -mouth-disease-exercise-blackthorn-evaluation-2018.pdf.

170 **sixty-four-page draft:** USDA, "Foot-and-Mouth Disease Resonse Plan: The Red Book," draft, October 2020, https://www.aphis.usda.gov/animal_health/emergency_ management/downloads/fmd_responseplan.pdf; and Agriculture Response Management and Resources (ARMAR), "Agriculture Response Management and Resources (ARMAR) Functional Exercise," June 29, 2018, https://www.ses-corp.com/2018/06/29 /agriculture-response-management-and-resources-armar-functional-exercise/.

171 **Saint-Hilaire's experiment:** Camille Limoges, "Zoological Adventures," *Science* (book review) 268, no. 5207 (April 7, 1995): 135–36; and Timothy Collins, "From Anatomy to Zoophagy: A Biographical Note on Frank Buckland on JSTOR," *Journal of the Galway Archaeological and Historical Society* 55 (2003): 91–109.

171 **"troops of elands":** Nicole Kearney, "Exploring the Acclimatisation Society of Victoria's Role in Australia's Ecological History," Biodiversity Heritage Library, January 25, 2018, accessed February 8, 2021, https://blog.biodiversitylibrary .org/2018/01/if-it-lives-we-want-it-exploring-the-acclimatisation-society-of -victorias-role-in-australias-ecological-history.html.

171 **society in Australia:** Acclimatization Society of Victoria, "Report of the Acclimatisation Society of Victoria," Google Books, accessed February 24, 2022, https://www .google.com/books/edition/Report_of_the_Acclimatisation_Society_of/_AcYAA AAYAAJ?hl=en&gbpv=1&dq=acclimatization+society+magpies+to+england&p g=RA1-PA28&printsec=frontcover.

172 **top one hundred globally invasive:** Global Invasive Species Database, "100 of the World's Worst Invasive Species," accessed April 26, 2022, http://www.iucngisd.org /gisd/100_worst.php.

172 **Lacey introduced:** Robert Anderson, "The Lacey Act: America's Premier Weapon in the Fight against Unlawful Wildlife Trafficking," *Public Land Law Review*, 16 Pub. L. L.R. 27 (1995), https://www.animallaw.info/article/lacey-act-americas-premier -weapon-fight-against-unlawful-wildlife-trafficking.

172 **"There is a compensation":** Theodore Whaley Cart, "The Lacey Act: America's First Nationwide Wildlife Statute," *Forest History Newsletter* 17, no. 3 (1973): 4–13.

173 **number of permits:** Susan D. Jewell, "A Century of Injurious Wildlife Listing under the Lacey Act: A History," *Management of Biological Invasions* 11, no. 3 (2020): 356–71.

173 **"No person shall":** Jewell, "A Century."

173 **the act prohibits:** Code of Federal Regulations, "CFR 16.13: Importation of Live or Dead Fish, Mollusks, and Crustaceans, or Their Eggs," accessed February 18, 2022, https://www.ecfr.gov/current/title-50/chapter-I/subchapter-B/part-16/subpart-B /section-16.13#p-16.13(e)(1).

174 **Defenders of Wildlife:** Defenders of Wildlife, "Petition to: Ken Salazar, Secretary, US Department of the Interior—Petition: To List All Live Amphibians in Trade as Injurious Unless Free of *Batrachochytrium dendrobatidis*," September 9, 2009, https://defenders .org/sites/default/files/publications/petition_to_interior_secretary_salazar.pdf.

174 **Lips, the coauthor:** Martel et al., "Recent Introduction of a Chytrid Fungus."

175 **some eight hundred thousand salamanders:** Yap et al., "Averting a North American Biodiversity Crisis."

175 **"We know what kind of killer":** Karen R. Lips and Joseph Mendelson III, "Stopping the Next Amphibian Apocalypse," *New York Times*, November 14, 2014.

176 **"renders Bsal introduction"**: Zhiyong Yuan et al., "Widespread Occurrence of an Emerging Fungal Pathogen in Heavily Traded Chinese Urodelan Species," *Conservation Letters* 11, no. 4 (July 1, 2018): e12436.

176 **native to Asia:** Yap et al., "Averting a North American Biodiversity Crisis."

176 **"Prior to the salamander rule":** Karen Lips, email communication with the author, January 31, 2021.

177 **large-scale monitoring might:** J. Hardin Waddle et al., "*Batrachochytrium salamandrivorans* (Bsal) Not Detected in an Intensive Survey of Wild North American Amphibians," *Scientific Reports* 10, 13012 (December 1, 2020).

177 **European Commission:** Frank Pasmans, email communication with the author, February 2021. See also "Commission Implementing Decision (EU) 2018/320," *Official Journal of the European Union*, February 28, 2018, https://eur-lex.europa.eu/legal-content/EN/TXT/PDF/?uri=CELEX:32018D0320&from=CS.

177 **"Whether traders really":** Pasmans, email communication.

177 **not worth the trouble:** Like in the United States, Bd is already endemic in the European Union and so, reasoned the European Commission, difficult to keep out. So far the fungus hasn't been as problematic in large parts of Europe because the circulating strains aren't all that virulent. This also makes the population vulnerable to highly virulent strains that may still come through with trade.

178 **"it's getting harder":** Faith Campbell, interview by the author, September 14, 2020.

178 **a small but targeted proportion:** USDA APHIS, "Risk-Based Sampling," last updated October 2, 2020, accessed April 26, 2022, https://www.aphis.usda.gov/aphis/ourfocus/planthealth/import-information/agriculture-quarantine-inspection/rbs.

178 **APHIS already requires:** USDA APHIS, "*Ralstonia*," last updated June 18, 2020, accessed January 3, 2022, https://www.aphis.usda.gov/aphis/ourfocus/planthealth/plant-pest-and-disease-programs/pests-and-diseases/plant-disease/sa_ralstonia/ct_ralstonia.

179 **"are grown by the gazillions":** Campbell, interview.

179 **"Just say no":** Campbell, interview.

179 *Phytophthora infestans*: This pathogen also causes blight in tomato plants. W. E. Fry et al., "The 2009 Late Blight Pandemic in the Eastern United States: Causes and Results," *Plant Disease* 97, no. 3 (2013): 296–306.

179 **One was Ohio:** Sara Hagan, "Tree-Killing Fungus Found in Ohio," *Journal News* (Butler County, Ohio), July 22, 2019.

180 **Rapid diagnostics:** Ganeshamoorthy Hariharan and Kandeeparoopan Prasannath, "Recent Advances in Molecular Diagnostics of Fungal Plant Pathogens: A Mini Review," *Frontiers in Cellular and Infection Microbiology*, 10:600234 (January 11, 2021).

181 **Antibiotic Resistance Lab Network:** Centers for Disease Control and Prevention, "About the AR Lab Network," accessed April 7, 2022, https://www.cdc.gov/drugresistance/laboratories.html.

181 **nursing homes:** Ellora Karmarkar et al., "LB1. Regional Assessment and Containment of *Candida auris* Transmission in Post-Acute Care Settings—Orange County, California, 2019," *Open Forum Infectious Diseases* 6, suppl. 2 (October 23, 2019): S993–S993.

181 *C. auris* **favors:** Matt Richtel, "With All Eyes on Covid-19, Drug-Resistant Infections Crept In," *New York Times*, January 27, 2021.

181 **acute care hospitals:** Centers for Disease Control and Prevention, "Fungal Diseases and COVID-19," accessed January 3, 2022, https://www.cdc.gov/fungal/covid-fungal.html; and C. Prestel et al., "*Candida auris* Outbreak in a COVID-19 Specialty Care Unit—Florida, July–August 2020," *Morbidity and Mortality Weekly Report* 70, no. 2 (January 15, 2021): 56–57.

181 **Before COVID-19:** Centers for Disease Control and Prevention, "Tracking *Candida auris*," accessed November 21, 2022, https://www.cdc.gov/fungal/candida-auris /tracking-c-auris.html; for more about the rise of hospital acquired infections during COVID-19 see: https://www.cdc.gov/hai/data/portal/covid-impact-hai.html.

182 **"For most of my life":** Tom Chiller, interview by the author, January 15, 2020.

182 **Valley fever:** Kerry Klein, "Valley Fever Could Spread with Climate Change, Study Warns," Valley Public Radio, KVPR, October 1, 2019, https:// www.kvpr.org/health/2019-10-01/valley-fever-could-spread-with-climate -change-study-warns#stream/0.

182 **"We really don't have":** Chiller, interview.

183 **"to improve the lives":** Day One Project, "About Us," accessed January 3, 2022, https://www.dayoneproject.org/about.

183 **Karen Lips submitted a proposal:** Karen R. Lips, "Improving Federal Management of Wildlife Movement and Emerging Infectious Disease," Day One Project, October 20, 2020, accessed June 13, 2022, https://www.dayoneproject.org/ideas/improving -federal-management-of-wildlife-movement-and-emerging-infectious-disease.

CHAPTER 10: RESPONSIBILITY

185 **a mold problem:** Natalia Novikova et al., "Survey of Environmental Biocontamination on Board the International Space Station," *Research in Microbiology* 157, no. 1 (2006): 5–12; and Novikova et al., "The Results of Microbiological Research of Environmental Microflora of Orbital Station *Mir* on Environmental Systems," in *SAE Technical 31st International Conference on Environmental Systems* (Orlando: Society of Automotive Engineers, 2001); "Mutant Fungus from Space," BBC News, March 8, 2001, accessed January 3, 2022, http://news.bbc.co.uk/2/hi/world/monitoring/ media_reports/1209034.stm.

185 **dozens of species:** Novikova et al., "Survey of Environmental Biocontamination."

185 **survive on the *outside*:** Marta Cortesão, interview by the author, September 23, 2020.

186 **October 1957:** Joshua Lederberg, "Exobiology: Approaches to Life beyond the Earth," *Science* 132, no. 3424 (January 3, 1960): 393–400.

186 **"history shows":** Lederberg, "Exobiology."

186 **Lederberg wondered:** Joshua Lederberg, "Can We Keep Mars Clean?" *Washington Post*, February 19, 1967, https://profiles.nlm.nih.gov/spotlight/bb/catalog/nlm :nlmuid-101584906X1086-doc.

187 **"Would we not deplore":** Lederberg, "Exobiology."

187 **"a moral disaster":** As quoted in Michael Meltzer, *When Biospheres Collide: A History of NASA's Planetary Protection Program* (Washington, DC: US Government Printing Office, 2011).

187 **"It would be rash":** Lederberg, "Exobiology."

187 **"the two big earthly superpowers":** Andy Spry, interview by the author, March 11, 2021.

187 **swabbed for microbes:** Moogega Cooper, "Planetary Protection: Protecting the Earth from the Universe . . . and the Universe from Earth," YouTube, NASA von Karman Lecture Series, February 4, 2021, https://www.youtube.com/watch?v =nPC1IJ5QgsA.

188 **"Special Regions":** John D. Rummel et al., "A New Analysis of Mars 'Special Regions': Findings of the Second MEPAG Special Regions Science Analysis Group (SR-SAG2)," *Astrobiology* 14, no. 11 (November 2014): 887–968.

188 **five hundred thousand spores:** Jet Propulsion Laboratory, NASA, "Mars 2020

Perseverance Launch Press Kit: Biological Cleanliness," accessed January 7, 2022, https://www.jpl.nasa.gov/news/press_kits/mars_2020/launch/mission/spacecraft /biological_cleanliness/.

188 **"Most organisms"**: Spry, interview.

189 **Marta Cortesão:** Marta Cortesão et al., "*Aspergillus niger* Spores Are Highly Resistant to Space Radiation," *Frontiers in Microbiology* 11:560. (April 2020).

189 **"For human health"**: Marta Cortesão, interview by the author, September 23, 2020.

189 **"completely destroyed"**: Cortesão, interview.

190 **some earthly organisms:** Y. Kawaguchi et al., "DNA Damage and Survival Time Course of Deinococcal Cell Pellets During 3 Years of Exposure to Outer Space," *Frontiers in Microbiology* 11 (2020): 2050, https://doi.org/10.3389/fmicb.2020.02050

190 **high doses:** Cortesão, "*Aspergillus niger*."

190 **"Layers are like"**: Cortesão, interview.

190 **"When you ask"**: Spry, email communication with the author, December 8, 2021.

191 **"Most scientists"**: "Space: Is the Earth Safe From Lunar Contamination?," *Time*, June 13, 1969, accessed May 1, 2022, http://content.time.com/time/subscriber/article /0,33009,942095-2,00.html.

191 **"All the stuff"**: "Charles A. Berry Oral History," interview by Carol Butler, Houston, TX, April 29, 1999, accessed May 1, 2022, https://historycollection.jsc.nasa.gov/JSC HistoryPortal/history/oral_histories/BerryCA/BerryCA_4-29-99.htm.

192 **Microbes could have infected:** Dagomar Degroot, "What Can We Learn from the Lunar Pandemic That Never Was?" *Aeon*, December 22, 2020, accessed January 7, 2022, https://aeon.co/essays/what-can-we-learn-from-the-lunar-pandemic-that-never -was.

192 **"A dangerous pathogen"**: Degroot, "What Can We Learn?"

192 **"For that, an infection needs"**: Andy Spry, email communication, May 1, 2022.

192 **before return to Earth:** NASA, "With First Martian Samples Packed, *Perseverance* Initiates Remarkable Sample Return Mission," October 12, 2021, accessed March 8, 2022, https://www.nasa.gov/feature/goddard/2021/with-first-martian-samples -packed-perseverance-initiates-remarkable-sample-return-mission.

193 **"viruses small and simple"**: Matthew Fisher, interview by the author, March 24, 2020.

194 **Hope remains:** John Messerly, "Hope and Pandora's Box," Reason and Meaning (blog), March 11, 2017, https://reasonandmeaning.com/2017/03/11/hope-and -pandoras-box/.

195 **"No one nation"**: Robert A. Cook, William Karesh, and Steven A. Osofsky, "Conference Summary, One World, One Health: Building Interdisciplinary Bridges to Health in a Globalized World," Rockefeller University, New York, September 29, 2004, http://www.oneworldonehealth.org/sept2004/owoh_sept04.html.

195 **"One Health"**: American Society for Microbiology, "One Health: Fungal Pathogens of Humans, Animals, and Plants," report on an American Academy of Microbiology Colloquium, Washington, DC, October 18, 2017, https://www.ncbi.nlm.nih.gov/books /NBK549988/.

195 **preserve life:** Rita Algorri, "A One Health Approach to Combating Fungal Disease: Forward-Reaching Recommendations for Raising Awareness," American Society for Microbiology, September 27, 2019, https://asm.org/Articles/2019/September/A -One-Health-Approach-to-Combating-Fungal-Disease.

FURTHER READING

FUNGI IN GENERAL

There are many good books about fungi. Here are a few:

E. C. Large. *The Advance of the Fungi.* New York: Henry Holt, 1940.
Nicholas P. Money. *The Triumph of the Fungi: A Rotten History.* Oxford: Oxford University Press, 2006.
Merlin Sheldrake. *Entangled Life: How Fungi Make Our Worlds, Change Our Minds and Shape Our Futures.* New York: Random House, 2020.
Suzanne Simard. *Finding the Mother Tree: Discovering the Wisdom of the Forest.* New York: Knopf, 2021.

FOOD

Daniel Charles. *Lords of the Harvest: Biotech, Big Money, and the Future of Food.* Cambridge, MA: Perseus, 2001.
Cary Fowler. *Seeds on Ice: Svalbard and the Global Seed Vault.* Westport, CT: Prospecta, 2016.
Dan Koeppel. *Banana: The Fate of the Fruit That Changed the World.* New York: Plume, 2008.
Charles C. Mann. *The Wizard and the Prophet: Two Remarkable Scientists and Their Dueling Visions to Shape Tomorrow's World.* New York: Vintage, 2018.
Stuart McCook. *Coffee Is Not Forever: A Global History of the Coffee Leaf Rust.* Athens: Ohio University Press, 2019.
Dan Saladino. *Eating to Extinction: The World's Rarest Foods and Why We Need to Save Them.* New York: Farrar, Straus and Giroux, 2022.
John Soluri. *Banana Cultures: Agriculture, Consumption and Environmental Change in Honduras and the United States.* Austin: University of Texas Press, 2005.
Daniel Stone. *The Food Explorer: The True Adventures of the Globe-Trotting Botanist Who Transformed What America Eats.* New York: Dutton, 2018.

OTHER SPECIES

Anne Biklé and David Montgomery. *The Hidden Half of Nature: The Microbial Roots of Life and Health.* New York: W. W. Norton, 2015.

Susan Freinkel. *American Chestnut: The Life, Death, and Rebirth of a Perfect Tree.* Berkeley: University of California Press, 2007.

Elizabeth Kolbert. *The Sixth Extinction: An Unnatural History.* New York: Henry Holt, 2014.

Michael Melzer. *When Biospheres Collide: A History of NASA's Planetary Protection Program.* Washington, DC: US Government Printing Office, 2011.

Donald Peattie. *A Natural History of North American Trees.* San Antonio, TX: Trinity University Press, 2013.

Diana Tomback, Stephen Arno, and Robert Keane, eds. *Whitebark Pine Communities: Ecology and Restoration.* Washington, DC: Island, 2001.

Jonathan Weiner. *The Beak of the Finch.* New York: Vintage, 1995.

If you would like to learn more or help prevent and resolve fungal pandemics, here are just a few nonprofits that support the work of scientists, physicians, policy makers, and farmers.

WILDLIFE

Amphibiaweb (https://amphibiaweb.org)
Bat Conservation International (https://www.batcon.org/)
Center for Biological Diversity (https://www.biologicaldiversity.org/)
Defenders of Wildlife (https://defenders.org/)
The Nature Conservancy (https://www.nature.org/en-us/)

HUMANS

Aspergillosis Trust (https://www.aspergillosistrust.org/)
Global Action for Fungal Infections (https://gaffi.org/)
ProMED (https://promedmail.org/)

TREES

Akaka Foundation for Tropical Forests (https://akakaforests.org)
American Chestnut Cooperators Foundation (https://www.accf-chestnut.org/)
The American Chestnut Foundation (acf.org)
The Sugar Pine Foundation (www.sugarpinefoundation.org)
Whitebark Pine Ecosystem Foundation (www.whitebarkfound.org and https://whitebark pine.ca/)

FOOD

Bananalink (bananalink.org.uk)

There are many seed banks large and small around the globe. Here are a few:

ASEED Europe (https://aseed.net/)
Biodiversity International and International Center for Tropical Agriculture (https://alliancebioversityciat.org/)
Crop Trust (Croptrust.org)

INDEX